In Silico Drug Design and Discovery: Big Data for Small Molecule Design 2nd Edition

In Silico Drug Design and Discovery: Big Data for Small Molecule Design 2nd Edition

Guest Editors

Carmen Cerchia
Antonio Lavecchia

MDPI

Basel • Beijing • Wuhan • Barcelona • Belgrade • Novi Sad • Cluj • Manchester

Guest Editors
Carmen Cerchia
Department of Pharmacy
University of Naples
Federico II
Naples
Italy

Antonio Lavecchia
Department of Pharmacy
University of Naples
Federico II
Naples
Italy

Editorial Office
MDPI AG
Grosspeteranlage 5
4052 Basel, Switzerland

This is a reprint of the Special Issue, published open access by the journal *Biomolecules* (ISSN 2218-273X), freely accessible at: https://www.mdpi.com/journal/biomolecules/special_issues/O9UPH2EJLK.

For citation purposes, cite each article independently as indicated on the article page online and as indicated below:

Lastname, A.A.; Lastname, B.B. Article Title. *Journal Name* **Year**, *Volume Number*, Page Range.

ISBN 978-3-7258-7839-0 (Hbk)
ISBN 978-3-7258-7840-6 (PDF)
https://doi.org/10.3390/books978-3-7258-7840-6

Contents

About the Editors

Carmen Cerchia

Carmen Cerchia is a fixed-term researcher at the Department of Pharmacy, University of Naples Federico II. She obtained her PhD in Pharmaceutical Sciences in 2016 under the supervision of Antonio Lavecchia and was then a visiting researcher for one year in Jürgen Bajorath's group at the Department of Life Science Informatics, B-IT, Rheinische Friedrich-Wilhelms-Universität Bonn. Her research focuses on computational chemistry approaches to drug discovery, including data mining and virtual screening applied to biological targets involved in cancer and metabolic diseases.

Antonio Lavecchia

Antonio Lavecchia is Professor of Medicinal Chemistry and Head of the Drug Discovery Laboratory at the Department of Pharmacy, University of Naples Federico II, where he is also the Scientist Responsible for the Molecular Modeling Excellence Laboratory. He received his PhD in Pharmaceutical Sciences from the University of Catania in 1999, with research conducted at the University of Minnesota. His work integrates computational chemistry with biological experiments, with a strong focus on artificial intelligence applications in drug discovery. His research has led to algorithms and protocols for the identification of novel drug candidates in cancer, diabetes, dyslipidemia, pain, and neurodegenerative disorders. He has authored more than 180 publications, three books, six book chapters, and six PCT patents, and has received several recognitions, including the Farmindustria Prize and Start Cup Campania awards. In 2021, Expertscape recognized him as a world expert in PPARs.

Preface

This Reprint was conceived to provide readers with a focused view of how computational chemistry and data-intensive methods are transforming drug design and discovery. The rapid expansion of bioactivity data, structural information, and machine learning tools has created new opportunities to address complex pharmacological questions with greater speed and depth. At the same time, the field is moving beyond isolated predictions toward integrated workflows in which modeling, simulation, data mining, and experimental validation contribute to a common decision-making process. The contributions presented here reflect this transition through studies on target characterization, virtual screening, safety assessment, affinity prediction, and generative design. We hope this Reprint will be useful to researchers in medicinal chemistry, chemical biology, bioinformatics, and related disciplines, as well as to students seeking an updated perspective on current in silico strategies for small-molecule discovery.

Carmen Cerchia and Antonio Lavecchia
Guest Editors

biomolecules

MDPI

Editorial

In Silico Drug Design and Discovery: Big Data for Small Molecule Design—2nd Edition

Carmen Cerchia * and Antonio Lavecchia

Drug Discovery Laboratory, Department of Pharmacy, University of Naples Federico II, 80131 Naples, Italy; antonio.lavecchia@unina.it

* Correspondence: carmen.cerchia@unina.it

The volume and heterogeneity of data available to drug discovery have grown at a pace that would have been difficult to predict even a few years ago. Public repositories of bioactivity measurements, crystallographic coordinates, and gene-expression profiles now collectively catalog billions of data points [1–12], while protein structure prediction has opened entire families of targets to computational interrogation [13–17]. At the same time, the way that this data is used has changed. Computational chemistry is no longer a discipline that operates in a silo, feeding one-off predictions into a linear pipeline; instead, it functions as the connective tissue of multidisciplinary discovery campaigns in which machine learning, physics-based simulation, and experimental feedback loops are interwoven from the outset [18–22]. This shift is reshaping the expectations placed on in silico methods [23]. Scale still matters, but it is no longer sufficient on its own. What distinguishes productive computational work today is the quality of the biological question asked, the rigor of the data curation underlying the models, and the willingness to close the loop with experiment. The contributions assembled in this second edition of the *Special Issue* "In Silico Drug Design and Discovery: Big Data for Small Molecule Design" illustrate this maturation across a remarkably diverse set of therapeutic areas and methodological strategies.

A first thematic thread running through the *Special Issue* concerns the use of structural modeling and simulation to illuminate poorly characterized targets. Adewumi and Mosebi tackled estrogen receptor-α36, a clinically relevant but structurally unresolved variant implicated in endocrine-resistant breast cancer [24]. By constructing a homology model [25] of the ligand-binding domain and performing molecular dynamics (MD) simulations coupled with MM/PBSA free-energy calculations, they provided a structural rationale for the preferential binding of broussoflavonol B to ER-α36 over ER-α66, offering a starting point for the design of selective inhibitors. In a different biological context, Giordano et al. combined sequence analysis, phylogenetics, homology modeling, and docking to characterize lipoxygenases from marine diatoms and to propose a classification linked to possible substrate preferences [26]. This study demonstrates how bioinformatic methods can convert sequence data into testable functional hypotheses. Rehman et al. applied a more classical discovery workflow to leucyl-tRNA synthetase from *Mycobacterium tuberculosis*, combining virtual screening with MD and then validating selected hits by isothermal titration calorimetry and antimycobacterial assays [27]. Their identification of Macimorelin as a promising binder of LeuRS underscores how computational repurposing screens, when reinforced by orthogonal experimental evidence, can yield viable chemical leads against drug-resistant pathogens. Yu et al. employed the SILCS methodology and MD to propose that bile acids and their conjugates act as negative allosteric modulators of the M1 muscarinic receptor, with potential implications for colon cancer biology [28]. This work brings together membrane-context signaling, endogenous metabolite pharmacology, and receptor allostery, three dimensions that are often studied in isolation.

A second theme is the transition from single-modality prediction toward integrated, data-rich models that capture multiple dimensions of molecular behavior [29,30]. Umemori et al. addressed a difficult problem in preclinical safety: the prediction of reactive metabolite formation associated with drug-induced liver injury [31,32], by developing a binary classifier for cysteine-trapping outcomes using proprietary experimental data [33]. Their head-to-head comparison of a message-passing neural network and a random forest model is instructive: under a realistic time-split validation, the graph-based architecture outperformed the fingerprint-based model, yet the analysis of contributory substructures retained interpretability in medicinal chemistry terms. This interplay between predictive accuracy and mechanistic transparency reflects a growing expectation in the field. Debnath et al. reported GramSeq-DTA, a framework for drug–target affinity prediction that combines structural representations of compounds and proteins with chemical perturbation signatures derived from the L1000 transcriptomic resource [34]. This study suggests that purely structure-based representations may not capture all the information relevant to affinity prediction. Piazza et al. provide a complementary illustration of how machine learning can be embedded into a practical discovery pipeline [35]. After developing and benchmarking a large panel of predictive models, they used the best performer to drive a virtual screening campaign for CDK9 inhibitors, ultimately identifying compounds with low-micromolar biochemical activity and promising cellular profiles. This work exemplifies how ML can complement rather than replace medicinal chemistry expertise, helping to narrow the candidate pool and focus experimental effort.

A distinct and increasingly important strand of contemporary computational drug design concerns generative models that learn the distribution of chemical space and propose novel molecules rather than simply scoring existing ones [36–41]. Islam and Caulfield developed a ligase-conditioned junction tree variational autoencoder to generate synthesizable, drug-like molecular glues aimed at promoting the degradation of amyloid-β42 through the ubiquitin–proteasome system [42]. This contribution is notable because it addresses targeted protein degradation in the context of neurodegeneration and treats generative design as a conditioned problem in which target biology, ligase specificity, and synthetic realism are jointly optimized.

Taken together, the contributions in this *Special Issue* suggest several directions that are likely to define the next phase of big-data-driven drug discovery. First, multimodality is becoming essential [43–45]. The most informative models will be those that integrate chemical structure, target structure, transcriptomic perturbation, safety endpoints, and real medicinal chemistry constraints within a single framework, rather than treating each data type separately [46,47]. Second, the coupling of data-driven and physics-based reasoning is gaining ground. Several papers in this *Special Issue* already combine predictive ranking with docking, MD, calorimetry, or cellular phenotyping, and such hybrid strategies are likely to become even more important as the community confronts sparse, noisy, and biased datasets [48]. Third, the standards by which computational work is evaluated are shifting. Retrospective benchmark performance alone is no longer a convincing endpoint; what matters is whether a model can deliver biologically meaningful insight and experimentally testable hypotheses with sufficient robustness [49–51]. Alongside these methodological trends, the community would benefit from greater attention to uncertainty quantification, data provenance, negative-result sharing, and model interpretability [52–55], elements that are still treated as optional refinements.

In conclusion, this second edition of "In Silico Drug Design and Discovery: Big Data for Small Molecule Design" offers a snapshot of a field that is maturing rapidly. The

 https://doi.org/10.3390/biom16040591

studies collected here indicate that the real promise of big data for small-molecule design is ultimately not automation, but the construction of closed-loop discovery systems in which computation, synthesis, and experiment iteratively refine one another, turning data into decisions, and decisions into better medicines.

Funding: This research received no external funding.

Acknowledgments: C.C. acknowledges MUR (Ministero dell'Università e della Ricerca), PON R&I 2014–2020 Asse IV "Istruzione e Ricerca per il recupero—REACT-EU", Azione IV.4—"Contratti di Ricerca su tematiche dell'Innovazione" and support from the University of Naples "Federico II" (Grant FRA 2022, Line C, INITIATIVE). During the preparation of this manuscript, the authors used ChatGPT (version 5.4, OpenAI) for the purposes of English language refinement. The authors have reviewed and edited the output and take full responsibility for the content of this publication.

Conflicts of Interest: The authors declare no conflicts of interest.

Abbreviations

The following abbreviations are used in this manuscript:

MD	Molecular Dynamics
SILCS	Site identification by ligand competitive saturation

References

1. Zdrazil, B. Fifteen Years of ChEMBL and Its Role in Cheminformatics and Drug Discovery. *J. Cheminform.* **2025**, *17*, 32. [CrossRef] [PubMed]
2. Kim, S.; Chen, J.; Cheng, T.; Gindulyte, A.; He, J.; He, S.; Li, Q.; Shoemaker, B.A.; Thiessen, P.A.; Yu, B.; et al. PubChem 2025 Update. *Nucleic Acids Res.* **2025**, *53*, D1516–D1525. [CrossRef] [PubMed]
3. Burley, S.K.; Bhatt, R.; Bhikadiya, C.; Bi, C.; Biester, A.; Biswas, P.; Bittrich, S.; Blaumann, S.; Brown, R.; Chao, H.; et al. Updated Resources for Exploring Experimentally-Determined PDB Structures and Computed Structure Models at the RCSB Protein Data Bank. *Nucleic Acids Res.* **2025**, *53*, D564–D574. [CrossRef] [PubMed]
4. Subramanian, A.; Narayan, R.; Corsello, S.M.; Peck, D.D.; Natoli, T.E.; Lu, X.; Gould, J.; Davis, J.F.; Tubelli, A.A.; Asiedu, J.K. A next Generation Connectivity Map: L1000 Platform and the First 1,000,000 Profiles. *Cell* **2017**, *171*, 1437–1452. [CrossRef]
5. Zhou, Y.; Zhang, Y.; Zhao, D.; Yu, X.; Shen, X.; Zhou, Y.; Wang, S.; Qiu, Y.; Chen, Y.; Zhu, F. TTD: Therapeutic Target Database Describing Target Druggability Information. *Nucleic Acids Res.* **2024**, *52*, D1465–D1477. [CrossRef]
6. Knox, C.; Wilson, M.; Klinger, C.M.; Franklin, M.; Oler, E.; Wilson, A.; Pon, A.; Cox, J.; Chin, N.E.; Strawbridge, S.A.; et al. DrugBank 6.0: The DrugBank Knowledgebase for 2024. *Nucleic Acids Res.* **2024**, *52*, D1265–D1275. [CrossRef]
7. Malik, A.; Arsalan, M.; Moreno, C.; Mosquera, J.; Félix, E.; Kizilören, T.; Muthukrishnan, V.; Zdrazil, B.; Leach, A.R.; O'Boyle, N.M. ChEBI: Re-Engineered for a Sustainable Future. *Nucleic Acids Res.* **2026**, *54*, D1768–D1778. [CrossRef]
8. The UniProt Consortium. UniProt: The Universal Protein Knowledgebase in 2025. *Nucleic Acids Res.* **2025**, *53*, D609–D617. [CrossRef]
9. Liu, T.; Hwang, L.; Burley, S.K.; Nitsche, C.I.; Southan, C.; Walters, W.P.; Gilson, M.K. BindingDB in 2024: A FAIR Knowledgebase of Protein-Small Molecule Binding Data. *Nucleic Acids Res.* **2025**, *53*, D1633–D1644. [CrossRef]
10. Béquignon, O.J.M.; Bongers, B.J.; Jespers, W.; IJzerman, A.P.; van der Water, B.; van Westen, G.J.P. Papyrus: A Large-Scale Curated Dataset Aimed at Bioactivity Predictions. *J. Cheminform.* **2023**, *15*, 3. [CrossRef]
11. Jiang, P.; Sinha, S.; Aldape, K.; Hannenhalli, S.; Sahinalp, C.; Ruppin, E. Big Data in Basic and Translational Cancer Research. *Nat. Rev. Cancer* **2022**, *22*, 625–639. [CrossRef] [PubMed]
12. Hunter, F.M.I.; Ioannidis, H.; Bento, A.P.; Bosc, N.; Corbett, S.; Felix, E.; Magarinos, M.P.; Manners, E.; Smit, I.A.; de Veij, M.; et al. Drug and Clinical Candidate Drug Data in ChEMBL. *J. Med. Chem.* **2025**, *68*, 19800–19827. [CrossRef] [PubMed]
13. Bertoni, D.; Tsenkov, M.; Magana, P.; Nair, S.; Pidruchna, I.; Querino Lima Afonso, M.; Midlik, A.; Paramval, U.; Lawal, D.; Tanweer, A.; et al. AlphaFold Protein Structure Database 2025: A Redesigned Interface and Updated Structural Coverage. *Nucleic Acids Res.* **2026**, *54*, D358–D362. [CrossRef] [PubMed]
14. Abramson, J.; Adler, J.; Dunger, J.; Evans, R.; Green, T.; Pritzel, A.; Ronneberger, O.; Willmore, L.; Ballard, A.J.; Bambrick, J.; et al. Accurate Structure Prediction of Biomolecular Interactions with AlphaFold 3. *Nature* **2024**, *630*, 493–500. [CrossRef]
15. Kovalevskiy, O.; Mateos-Garcia, J.; Tunyasuvunakool, K. AlphaFold Two Years on: Validation and Impact. *Proc. Natl. Acad. Sci. USA* **2024**, *121*, e2315002121. [CrossRef]

 https://doi.org/10.3390/biom16040591

16. Varadi, M.; Velankar, S. The Impact of AlphaFold Protein Structure Database on the Fields of Life Sciences. *Proteomics* **2023**, *23*, 2200128. [CrossRef]
17. Percudani, R.; De Rito, C. Predicting Protein Function in the AI and Big Data Era. *Biochemistry* **2025**, *64*, 2345–2352. [CrossRef]
18. Bienstock, R.J. AI/ML Methodologies and the Future-Will They Be Successful in Designing the next Generation of New Chemical Entities? *J. Cheminform.* **2025**, *17*, 46. [CrossRef]
19. Vamathevan, J.; Clark, D.; Czodrowski, P.; Dunham, I.; Ferran, E.; Lee, G.; Li, B.; Madabhushi, A.; Shah, P.; Spitzer, M.; et al. Applications of Machine Learning in Drug Discovery and Development. *Nat. Rev. Drug Discov.* **2019**, *18*, 463–477. [CrossRef]
20. Klambauer, G.; Hochreiter, S.; Rarey, M. Machine Learning in Drug Discovery. *J. Chem. Inf. Model.* **2019**, *59*, 945–946. [CrossRef]
21. Chen, W.; Liu, X.; Zhang, S.; Chen, S. Artificial Intelligence for Drug Discovery: Resources, Methods, and Applications. *Mol. Ther. Nucleic Acids* **2023**, *31*, 691–702. [CrossRef] [PubMed]
22. Deng, J.; Yang, Z.; Ojima, I.; Samaras, D.; Wang, F. Artificial Intelligence in Drug Discovery: Applications and Techniques. *Brief. Bioinform.* **2022**, *23*, bbab430. [CrossRef] [PubMed]
23. Zitnik, M. AI-Enabled Drug Discovery Reaches Clinical Milestone. *Nat. Med.* **2025**, *31*, 2490–2491. [CrossRef] [PubMed]
24. Adewumi, A.T.; Mosebi, S. Characteristic Binding Landscape of Estrogen Receptor-A36 Protein Enhances Promising Cancer Drug Design. *Biomolecules* **2023**, *13*, 1798. [CrossRef]
25. Nema, P.; Purohit, A.; Soni, V.; Kashaw, S.K. A Comprehensive Review on Homology Modelling in Drug Development: Methodologies, Tools, and Evaluation. *Mini Rev. Med. Chem.* **2026**, *26*. [CrossRef]
26. Giordano, D.; Bonora, S.; D'Orsi, I.; D'Alelio, D.; Facchiano, A. Structural and Functional Characterization of Lipoxygenases from Diatoms by Bioinformatics and Modelling Studies. *Biomolecules* **2024**, *14*, 276. [CrossRef]
27. Rehman, Z.U.; Najmi, A.; Zoghebi, K. Insights into the Effects of Ligand Binding on Leucyl-tRNA Synthetase Inhibitors for Tuberculosis: In Silico Analysis and Isothermal Titration Calorimetry Validation. *Biomolecules* **2024**, *14*, 711. [CrossRef]
28. Yu, W.; MacKerell, A.D.; Weber, D.J.; Raufman, J.-P. Bile Acids Are Potential Negative Allosteric Modulators of M1 Muscarinic Receptors. *Biomolecules* **2025**, *15*, 1326. [CrossRef]
29. Angelo, J.S.; Guedes, I.A.; Barbosa, H.J.C.; Dardenne, L.E. Multi-and Many-Objective Optimization: Present and Future in de Novo Drug Design. *Front. Chem.* **2023**, *11*, 1288626. [CrossRef]
30. Luukkonen, S.; van den Maagdenberg, H.W.; Emmerich, M.T.M.; van Westen, G.J.P. Artificial Intelligence in Multi-Objective Drug Design. *Curr. Opin. Struct. Biol.* **2023**, *79*, 102537. [CrossRef]
31. Park, B.K.; Boobis, A.; Clarke, S.; Goldring, C.E.P.; Jones, D.; Kenna, J.G.; Lambert, C.; Laverty, H.G.; Naisbitt, D.J.; Nelson, S.; et al. Managing the Challenge of Chemically Reactive Metabolites in Drug Development. *Nat. Rev. Drug Discov.* **2011**, *10*, 292–306. [CrossRef] [PubMed]
32. Ray, A.S. Practical Perspectives on Addressing Hepatotoxicity during Clinical Candidate Selection. *Expert Opin. Drug Metab. Toxicol.* **2025**, *21*, 751–753. [CrossRef] [PubMed]
33. Umemori, Y.; Handa, K.; Yoshimura, S.; Kageyama, M.; Iijima, T. Development of a Novel In Silico Classification Model to Assess Reactive Metabolite Formation in the Cysteine Trapping Assay and Investigation of Important Substructures. *Biomolecules* **2024**, *14*, 535. [CrossRef] [PubMed]
34. Debnath, K.; Rana, P.; Ghosh, P. GramSeq-DTA: A Grammar-Based Drug–Target Affinity Prediction Approach Fusing Gene Expression Information. *Biomolecules* **2025**, *15*, 405. [CrossRef]
35. Piazza, L.; Poles, C.; Bononi, G.; Granchi, C.; Di Stefano, M.; Poli, G.; Giordano, A.; Medugno, A.; Napolitano, G.M.; Tuccinardi, T.; et al. Machine Learning-Based Virtual Screening for the Identification of Novel CDK-9 Inhibitors. *Biomolecules* **2026**, *16*, 12. [CrossRef]
36. Cerchia, C.; Delre, P.; Falbo, E.; Lavecchia, A. Artificial Intelligence in De Novo Drug Design. In *Applied Artificial Intelligence for Drug Discovery: From Data-Driven Insights to Therapeutic Innovation*; Lavecchia, A., Ed.; Springer Nature: Cham, Switzerland, 2026; pp. 157–186, ISBN 978-3-031-98022-0.
37. Bilodeau, C.; Jin, W.; Jaakkola, T.; Barzilay, R.; Jensen, K.F. Generative Models for Molecular Discovery: Recent Advances and Challenges. *Wiley Interdiscip. Rev. Comput. Mol. Sci.* **2022**, *12*, e1608. [CrossRef]
38. Cheng, Y.; Gong, Y.; Liu, Y.; Song, B.; Zou, Q. Molecular Design in Drug Discovery: A Comprehensive Review of Deep Generative Models. *Brief. Bioinform.* **2021**, *22*, bbab344. [CrossRef]
39. Tong, X.; Liu, X.; Tan, X.; Li, X.; Jiang, J.; Xiong, Z.; Xu, T.; Jiang, H.; Qiao, N.; Zheng, M. Generative Models for de Novo Drug Design. *J. Med. Chem.* **2021**, *64*, 14011–14027. [CrossRef]
40. Bajorath, J. Chemical Language Models for Molecular Design. *Mol. Inform.* **2024**, *43*, e202300288. [CrossRef]
41. Bhattacharya, D.; Cassady, H.J.; Hickner, M.A.; Reinhart, W.F. Large Language Models as Molecular Design Engines. *J. Chem. Inf. Model.* **2024**, *64*, 7086–7096. [CrossRef]
42. Islam, N.N.; Caulfield, T.R. Conditioned Generative Modeling of Molecular Glues: A Realistic AI Approach for Synthesizable Drug-like Molecules. *Biomolecules* **2025**, *15*, 849. [CrossRef] [PubMed]

43. Chakraborty, C.; Bhattacharya, M.; Pal, S.; Chatterjee, S.; Das, A.; Lee, S.-S. AI-Enabled Language Models (LMs) to Large Language Models (LLMs) and Multimodal Large Language Models (MLLMs) in Drug Discovery and Development. *J. Adv. Res.* **2025**, *78*, 377–389. [CrossRef] [PubMed]
44. Luo, Y.; Liu, X.Y.; Yang, K.; Huang, K.; Hong, M.; Zhang, J.; Wu, Y.; Nie, Z. Toward Unified Ai Drug Discovery with Multimodal Knowledge. *Health Data Sci.* **2024**, *4*, 0113. [CrossRef] [PubMed]
45. Elemento, O.; Leslie, C.; Lundin, J.; Tourassi, G. Artificial Intelligence in Cancer Research, Diagnosis and Therapy. *Nat. Rev. Cancer* **2021**, *21*, 747–752. [CrossRef]
46. Huynh, D.L.; Seal, S.; Reid, D.; Carpenter, A.E.; Bender, A.; Spjuth, O.; Seal, S.; Huynh, D.L.; Chelbi, M.; Patra, A.; et al. AI Agents in Drug Discovery: Applications and Case Studies. *Drug Discov. Today* **2026**, 104650. [CrossRef]
47. Markossian, S.; Baskir, H.M.; Grossman, A.C.; Chandran, S.; Rele, S.; Hall, M.D.; Amaro, R.E.; Tropsha, A.; Zakharov, A.V. In Silico Drug Discovery: Bridging the Gaps in Preclinical Translation. *Drug Discov. Today* **2026**, *31*, 104567. [CrossRef]
48. Fromer, J.C.; Coley, C.W. Computer-Aided Multi-Objective Optimization in Small Molecule Discovery. *Patterns* **2023**, *4*, 100678. [CrossRef]
49. Pitt, W.R.; Bentley, J.; Boldron, C.; Colliandre, L.; Esposito, C.; Frush, E.H.; Kopec, J.; Labouille, S.; Meneyrol, J.; Pardoe, D.A.; et al. Real-World Applications and Experiences of AI/ML Deployment for Drug Discovery. *J. Med. Chem.* **2025**, *68*, 851–859. [CrossRef]
50. Wang, Y.; Hu, Z.; Chang, J.; Yu, B. Thinking on the Use of Artificial Intelligence in Drug Discovery. *J. Med. Chem.* **2025**, *68*, 4996–4999. [CrossRef]
51. Doron, G.; Genway, S.; Roberts, M.; Jasti, S. Generative AI: Driving Productivity and Scientific Breakthroughs in Pharmaceutical R&D. *Drug Discov. Today* **2025**, *30*, 104272. [CrossRef]
52. Gangwal, A.; Lavecchia, A. IMPACT Framework: Establishing Global Standards for Artificial Intelligence Implementation, Methodology, and Translation in Drug Discovery. *WIREs Comput. Mol. Sci.* **2026**, *16*, e70072. [CrossRef]
53. Jiménez-Luna, J.; Grisoni, F.; Schneider, G. Drug Discovery with Explainable Artificial Intelligence. *Nat. Mach. Intell.* **2020**, *2*, 573–584. [CrossRef]
54. Lamens, A.; Bajorath, J. Explainable Artificial Intelligence for Molecular Design in Pharmaceutical Research. *Chem. Sci.* **2026**, *17*, 1411–1422. [CrossRef]
55. Bajorath, J. From Explainable Artificial Intelligence to Human Understanding. *Artif. Intell. Life Sci.* **2025**, *7*, 100131. [CrossRef]

https://doi.org/10.3390/biom16040591

biomolecules

MDPI

Article

Machine Learning-Based Virtual Screening for the Identification of Novel CDK-9 Inhibitors

Lisa Piazza [1], Clarissa Poles [2,3], Giulia Bononi [1], Carlotta Granchi [1], Miriana Di Stefano [1], Giulio Poli [1], Antonio Giordano [4], Annamaria Medugno [5], Giuseppe Maria Napolitano [6,*], Tiziano Tuccinardi [1,7,*] and Luigi Alfano [8]

1 Department of Pharmacy, University of Pisa, 56126 Pisa, Italy; lisa.piazza@phd.unipi.it (L.P.); giulia.bononi@unipi.it (G.B.); carlotta.granchi@unipi.it (C.G.); miriana.distefano@farm.unipi.it (M.D.S.); giulio.poli@unipi.it (G.P.)
2 Genomics and Experimental Medicine Program, Scuola Superiore Meridionale (SSM, School of Advanced Studies), Via Mezzocannone 4, 80078 Napoli, Italy; c.poles@tigem.it
3 Telethon Institute of Genetics and Medicine, Via Campi Flegrei 34, 80078 Napoli, Italy
4 Sbarro Institute for Cancer Research and Molecular Medicine, Center for Biotechnology, College of Science and Technology, Temple University, Philadelphia, PA 19122, USA; antonio.giordano@temple.edu
5 Department of Medical Biotechnologies, University of Siena, 53100 Siena, Italy; a.medugno1@studenti.unisi.it
6 Clinical and Translational Oncology Program, Scuola Superiore Meridionale (SSM, School of Advanced Studies), University of Naples Federico II, 80131 Napoli, Italy
7 Consorzio Interuniversitario Nazionale per la Scienza e Tecnologia dei Materiali (INSTM), 50121 Firenze, Italy
8 Department of Breast and Thoracic Oncology, Istituto Nazionale Tumori-IRCCS-Fondazione G. Pascale, 80131 Napoli, Italy; l.alfano@istitutotumori.na.it
* Correspondence: g.napolitano@ssmeridionale.it (G.M.N.); tiziano.tuccinardi@unipi.it (T.T.)

Abstract

Cyclin-dependent kinase 9 (CDK9) is a key regulator of transcriptional elongation and DNA repair, supporting cancer cell survival by sustaining the expression of oncogenes and anti-apoptotic proteins. Its overexpression in multiple malignancies makes it an attractive target for anticancer therapy. Here, we report a machine learning (ML) based approach to identify novel CDK9 inhibitors. Through systematic data collection and preprocessing, seventy predictive models were developed using five algorithms, two classification settings, and seven molecular representations. The best-performing model was employed to guide a virtual screening (VS) campaign, resulting in the identification of 14 compounds promising for their potential inhibitory effect. Upon enzymatic assays, two molecules with inhibitory activity in the low micromolar range were selected as promising candidates and further tested in three cancer cell lines with distinct genetic backgrounds. These experiments led to the identification of a novel compound exhibiting interesting therapeutic potential, both as a single agent and in combination with Camptothecin (CPT), revealing varying response profiles across the tested cell lines. These results illustrate the power of integrating ML within anticancer drug discovery pipelines and represent a valuable starting point for the development of novel CDK9 inhibitors.

Keywords: cyclin-dependent kinase 9; machine learning; cancer therapy; drug discovery; virtual screening

1. Introduction

Protein kinases are essential regulators of intracellular signaling, controlling a broad spectrum of cellular processes, including proliferation, differentiation, and apoptosis, through the phosphorylation of specific substrates. Cyclin-dependent kinases (CDKs) form a well-characterized subfamily of serine/threonine kinases, whose activation is tightly regulated by the binding of specific cyclin proteins. CDKs are classically divided into two functional groups: those primarily involved in cell cycle control (e.g., CDK1, CDK2, CDK4, and CDK6), and those that regulate gene transcription through RNA polymerase II (e.g., CDK7, CDK8, CDK9, CDK12, CDK13, and CDK19) [1]. More recently, the role of transcription-associated CDKs in oncogenesis has gained increasing attention. Among these, CDK9 is a multifunctional kinase regulating many different processes from DNA repair to transcriptional elongation [2]. In complex with cyclin T1, it forms the positive transcription elongation factor b (P-TEFb), which phosphorylates the C-terminal domain (CTD) of RNA polymerase II, as well as the elongation factors DSIF (DRB sensitivity-inducing factor) and NELF (negative elongation factor). This phosphorylation event is crucial for overcoming the promoter-proximal pausing that occurs shortly after transcription initiation, a process known as the early elongation checkpoint. Without CDK9 activity, RNA polymerase II remains stalled, and productive elongation is impaired, resulting in incomplete transcript synthesis [1,3,4]. CDK9 activation itself depends on phosphorylation of a conserved threonine residue (T186) within the T-loop, mediated by autophosphorylation or upstream kinases such as CDK7, which stabilizes the active conformation of the enzyme [4].

Given this central role in transcriptional control, CDK9 is frequently exploited by cancer cells to sustain the expression of oncogenes and short-lived anti-apoptotic proteins such as MYC and MCL-1. Its overexpression has been observed in several solid and hematological malignancies, including pancreatic, ovarian, endometrial, and colorectal cancers, where it is associated with poor prognosis, enhanced metastatic potential, and resistance to conventional therapies [5,6]. In addition to the predominant exploration for cancer treatment, CDK9 inhibition has gained relevance in a broader spectrum of medical areas suffering significant unmet needs, such as autoimmune, inflammatory, and cardiac diseases [7,8], further underscoring the therapeutic promise of CDK9.

Nevertheless, a straightforward clinical translation remains hampered by a narrow therapeutic window and poor clinical response [9]. These issues strengthen the need for fast and advanced approaches to identify novel CDK9 inhibitors with improved pharmacological profiles.

In recent years, computational approaches have emerged as powerful tools to complement traditional drug discovery. In silico methods such as molecular docking, pharmacophore modeling, and machine learning (ML) offer efficient means to predict compound–target interactions, reduce screening costs, and accelerate lead identification. Among these, ML techniques have demonstrated notable success in modeling complex structure-activity relationships, enabling the prioritization of promising molecules as novel potent kinase inhibitors from large chemical libraries with increased accuracy [10–12]. In this study, we applied an ML-driven virtual screening (VS) strategy to identify novel small molecules acting as inhibitors of CDK9. Curated bioactivity datasets, molecular fingerprints, and advanced classification models were integrated to predict inhibitory potential. Top-ranking candidates resulting from VS were first selected and subjected to enzymatic assays to experimentally validate their predicted inhibitory activity toward CDK9. Among them, the two most promising compounds were further evaluated in cell-based assays, leading to the identification of a novel CDK9 inhibitor displaying significant cytotoxic activity. Finally, this compound was investigated through molecular docking and molecular dynamics (MD) simulations to characterize its interaction with CDK9, and to propose a plausible binding mode within the protein active site. Notably, this analysis also represents a valuable contribution to support future CDK9 studies by providing a solid structural rationale for subsequent optimization efforts. To further enhance the practical usability of the developed model, we made it publicly available through a dedicated GitHub repository

https://doi.org/10.3390/biom16010012

(https://github.com/MMVSL/CDK9Screen, accessed on 25 October 2025), thereby providing the scientific community with a transparent, accessible, and easy-to-use computational tool to predict the potential inhibitory activity of candidates of interest against CDK9.

2. Materials and Methods

2.1. Data Processing

The ChEMBL database [13] (version 31) was chosen as the primary activity data repository. Compounds with reported inhibitory activity against CDK9, either as a single protein (target ChEMBL ID: CHEMBL3116, 2213 entries) or complexed with cyclin T1 (target ChEMBL ID: CHEMBL2111389, 2380 entries), were retrieved and processed through the following filtering pipeline. Firstly, only activity records with a standard relation type of "=" (i.e., exact values) for IC_{50}, EC_{50}, K_i, or K_d were retained. Entries with missing SMILES (Simplified Molecular Input Line Entry System) or unspecified activity values were discarded. Since IC_{50} data represented approximately 80% of the available measurements, we decided at this stage to prioritize these entries for model development, thus using them as training compounds, while reserving records with EC_{50}, K_i, and K_d values as test set compounds for model evaluation. In light of this, when multiple measurements were available for the same compound, the activity expressed as IC_{50} was preferred in order to enrich the training set. This procedure resulted in two preliminary datasets comprising 1914 compounds for the training set and 464 for the test set.

Subsequently, both the preliminary training and test sets underwent duplicate instance removal and structural standardization, including salt elimination and ionization at physiological pH using the software MolBook Pro [14,15]. This preprocessing pipeline resulted in the identification of 1629 final compounds designated for the training set (the distribution of their activity values is reported in Figure S1) and 372 compounds for the test set.

Two classification approaches were then defined to distinguish active from inactive compounds, resulting in two separate classification schemes. Each approach was applied independently to both the training and test sets, generating a total of four datasets, one classified train/test pair for each classification scheme. According to the first approach (Lead-Oriented, LO), compounds with activity ≤ 1 μM, which corresponds to p*Activity* ≥ 6, were labeled as active, while those with an activity ≥ 5 μM were labeled as inactive (p*Activity* ≤ 5.30). These thresholds were derived from a prior study conducted on CDK5, a kinase from the same protein family for which this cutoff proved to be effective [11].

A second classification scheme (Potency-Oriented, PO) was defined to emphasize compounds with higher potency. In this case, molecules with activity ≤ 31.6 nM, which corresponds to p*Activity* ≥ 7.5, were labeled as active, while those with activity ≥ 3.2 μM, corresponding to p*Activity* ≤ 5.5, were categorized as inactive. In each classification scheme, compounds with activity values falling within the two classification thresholds were excluded from the corresponding dataset.

As a result, the classified training set for the PO scheme contained 565 molecules, of which 447 were classified as active and 118 as inactive. The training set of the LO approach instead included a total of 1465 compounds (1375 classified as active; 90 classified as inactive). Similarly, the corresponding test sets comprised 150 compounds for the PO scheme (106 labelled as active and 44 labelled as inactive) and 309 compounds for the LO scheme (279 marked as active; 30 marked as inactive).

Subsequently, to mitigate the pronounced imbalance toward the active class observed across all four datasets, additional inactive compounds were incorporated from the ChEMBL 31 database. These entries corresponded to records in which the reported

 https://doi.org/10.3390/biom16010012

activity was expressed with the relation symbol ">", meaning that the actual activity value was higher than the one reported. During the initial data processing, such compounds had been excluded because this format did not provide a precise numerical value. Once the activity thresholds were established, however, compounds reported with a ">" relation were confidently assigned to the inactive class whenever their recorded activity exceeded the predefined inactivity cutoff. Precisely, 228 and 62 inactive compounds were added to the training and test sets of the PO approach, respectively.

Within the LO scheme, instead, this step resulted in the integration of 224 inactive molecules in the training set, and 246 in the test set. Despite this first integration, the datasets remained unbalanced. Therefore, an additional set of inactive decoys was retrieved from the ZINC20 database [16], as previously performed [12]. All decoy compounds underwent the same data processing and standardization protocol described previously to guarantee consistency across the integrated datasets. Additionally, a control procedure was performed to ensure that no instances were in common between each train and the corresponding test set, preserving the integrity of the model evaluation process. As a result, four curated datasets were obtained, each perfectly balanced in terms of class distribution; i.e., LO training and test sets containing 2750 and 558 compounds, respectively, and the corresponding PO datasets with 894 (training) and 212 (test) instances.

2.2. Molecular Representations

To enable ML algorithms to interpret chemical structures and detect relevant features to make accurate predictions, molecules must first be translated into numerical representations. In this study, we explored seven types of molecular encodings: three established fingerprints (RDKit, Morgan, and PubChem fingerprint), a set of 208 numerical descriptors derived using RDKit, and three novel hybrid representations obtained by the concatenation of each fingerprint with descriptors. All molecular representations were generated starting from the SMILES strings of each compound, exploiting functionalities included in the RDKit Python library (version 2024.3.5) [17].

RDKit fingerprint: this representation captures molecular structure by identifying various connectivity patterns, known as subgraphs, that correspond to atom-bond paths of different lengths within a molecule. Each subgraph is then encoded into a specific position in a binary vector through a hashing process. RDKit fingerprints were generated using the RDKit library with a vector length of 1024.

Morgan fingerprint: this method encodes structural information by iteratively analyzing the neighborhood of each heavy atom up to a defined radius. In this study, we set a radius of two bonds. Each local atomic environment is then translated into a numerical identifier through a hashing technique. Coherently with RDKit fingerprints, Morgan fingerprints were computed with a fixed vector length of 1024 bits using the RDKit toolkit.

PubChem fingerprint: this method employs a predefined dictionary of chemical features. Each molecule is screened for the presence of 881 specific substructures, with a corresponding binary vector used to record their presence (1) or absence (0).

RDKit Descriptors: these numerical features reflect various chemical and physical properties of a molecule. Using RDKit's built-in descriptor functions, we calculated 208 attributes per compound and normalized them using Scikit-learn Min-Max scaler [18] to ensure uniform magnitude across the dataset.

Hybrid representations: these representations were generated by concatenating the descriptors with the Morgan (HybridMR), PubChem (HybridPR), or RDKit (HybridRR) fingerprint, respectively.

https://doi.org/10.3390/biom16010012

2.3. Machine Learning Algorithms

ML models were developed using two primary Python frameworks: Scikit-learn [18] for traditional algorithms and TensorFlow for neural network-based approaches. All models adopted a supervised classification strategy, designed to discriminate between compounds with potential inhibitory activity against CDK9 and those considered inactive. The output of each model consists of a probability score (PS) indicating the likelihood that a given compound functions as an active inhibitor. To build the classification models, five distinct algorithms were employed: Random Forest (RF), Support Vector Machine (SVM), k-Nearest Neighbors (KNN), Gaussian Process (GP), and Multi-Layer Perceptron (MLP). For each algorithm, different hyperparameters were optimized to find the most robust architecture setup.

RF combines a multitude of decision trees, each trained on a random subset of the training data sampled with replacement. The predictions of all trees are eventually aggregated through majority voting to determine the final output. The most relevant hyperparameters optimized for RF models were *n_estimators*, which specifies the total number of decision trees in the ensemble (with tested values 100 and 500), and *max_features*, which defines how many features are randomly selected at each node to identify the best split (with tested values *None*, *sqrt*, and *log2*).

SVM aims to identify an optimal hyperplane that best separates the two classes in a high-dimensional feature space. This space is implicitly defined through kernel functions, which enable the algorithm to compute inner products in the feature space without explicitly performing the transformation, an approach known as the kernel trick. As a result, all computations are carried out in the input space, making the method both efficient and scalable to complex, non-linear classification problems. The kernel functions explored during hyperparameter tuning were Scikit-learn available kernels such as *linear*, *poly*, *rbf*, *sigmoid*, and *Tanimoto* [19].

KNN classifies data points based on the majority label among the k closest training instances, using a distance metric to determine proximity. The number of neighbors to consider (with tested values being odd integers ranging from 3 to 15) and the option to weight neighbors based on their distance to the query compounds were included as tunable hyperparameters.

GP represents a non-parametric probabilistic framework that uses kernel functions to capture complex patterns in the data and estimate the likelihood that a sample belongs to a certain class. Different Scikit-learn kernel functions were evaluated during optimization, including *DotProduct*, *RationalQuadratic*, *Matern*, *WhiteKernel*, and *RBF*.

MLP is part of the deep learning family. This architecture consists of an input layer, one or more hidden layers, and an output layer. Each hidden layer contains multiple neurons, where each neuron applies a weighted sum of its inputs followed by a non-linear activation function, such as ReLU. This setup allows MLP to model complex, non-linear relationships within the data. The hidden layer size and the batch size were among the main optimized hyperparameters. For the batch size, the tested values were 100 and 200. For the hidden layer size, instead, all the possible combinations between (2, 3) layers and (100, 200, 1000) neurons in each layer were tested.

2.4. Model Evaluation and Structural Similarity Metrics

To evaluate model performance, we focused on three evaluation metrics: Matthews correlation coefficient (MCC) [20], precision, and recall. Their formulas are herein reported.

https://doi.org/10.3390/biom16010012

$$MCC = \frac{(TP \cdot TN) - (FP \cdot FN)}{\sqrt{(TP + FP)(TP + FN)(TN + FP)(TN + FN)}}$$

$$Precision = \frac{TP}{TP + FP}$$

$$Recall = \frac{TP}{TP + FN}$$

True Positives (TP) and true negatives (TN) denote the correctly predicted active and inactive compounds, respectively. In contrast, false positives (FP) are inactive molecules that the model incorrectly identifies as active, while false negatives (FN) are active compounds that the model fails to recognize, misclassifying them as inactive.

MCC provides a comprehensive measure of classification quality by incorporating all four confusion matrix outcomes. It ranges from −1 to 1, where 1 indicates perfect prediction accuracy with no FP or FN, and −1 corresponds to completely incorrect predictions. Precision measures the proportion of correctly predicted active compounds among all predicted actives, ranging from 0 to 1, with higher values indicating fewer false positives. It reflects the reliability of positive predictions but does not account for false negatives. On the other hand, recall quantifies the model's ability to identify all true active compounds, and it also ranges from 0 to 1.

In addition to these evaluation metrics, the Tanimoto coefficient [21] was used to assess molecular similarity. This measure compares molecular fingerprints of two compounds and yields values between 0 and 1, with values closer to 1 indicating higher structural similarity. The formula for Tanimoto similarity is herein reported:

$$T = \frac{s}{a + b - s}$$

where s is the number of shared bits set to 1 in both fingerprints, while a and b represent the total number of bits set to 1 in the fingerprint of the first and second molecule, respectively.

2.5. Model Development and Assessment

For each classification scheme, 35 ML models were derived by systematically combining 7 molecular representations with 5 ML algorithms. For classical ML approaches, namely RF, SVM, GP, and KNN, each model underwent a hyperparameter optimization phase using Scikit-learn's Grid Search [18]. This method systematically explores predefined hyperparameter combinations through cross-validation (CV) to identify the best configuration based on the MCC, used here as the primary performance metric. Following tuning, each model was further evaluated through 5-fold CV to ensure robustness and consistency across different data splits. For every fold, the training data were randomly partitioned into 80% for model training and 20% for internal validation. An analogous workflow was applied to MLP-based models using a custom pipeline implemented in TensorFlow, adapting hyperparameter tuning and validation steps to neural network architectures.

Once the optimal parameters were identified, the final models were retrained on the complete training set and evaluated against the independent test set to assess their generalization capability.

All training procedures were executed on a workstation equipped with an Intel Core i7-12700K CPU, using 16 cores for parallelized operations where supported.

2.6. Virtual Screening

The VS library was generated starting from the Aldrich Market Select (AMS) 2022 catalogue. Compounds were curated using a standardization protocol consistent with

 https://doi.org/10.3390/biom16010012

the data processing steps applied to training and test sets, including salt removal and ionization at physiological pH. As a result, we obtained a final VS dataset containing a total of 7,722,826 commercial and properly standardized compounds.

The VS was performed on a workstation equipped with an Intel Core i7-12700K CPU.

2.7. Clustering

To select a structurally diverse subset of high-confidence candidates for in vitro testing, hierarchical agglomerative clustering was applied to the VS hits using Orange3 software version 3.36.0 [22]. In this method, each compound initially constitutes a separate cluster. Clusters are then iteratively merged based on pairwise distances, progressively assembling the most similar ones until all compounds are integrated into a single hierarchical structure. Compounds were encoded as RDKit fingerprints, and the similarity between compounds was measured using the Tanimoto coefficient computed via the RDKit package version 2024.3.5. Distance was then defined as one minus the Tanimoto similarity. Clusters were merged based on the average linkage criterion, which defines the distance between two clusters as the mean pairwise distance between their members.

2.8. Experimental Validation

2.8.1. Enzymatic Assays

The inhibitory activity of the compounds identified by the VS was evaluated against CDK9/cyclin T1 using the fluorescence-based ADAPTA kinase assay (Thermo Fisher Scientific, Madison, WI, USA). The assay quantifies ADP generated during the kinase reaction through a competitive immunoassay employing a TR-FRET detection system. In the presence of an inhibitor, ADP formation by the kinase is reduced, leading to preservation of the antibody–tracer complex and a corresponding increase in the TR-FRET signal. Inhibitory potency was expressed as IC_{50} values, determined from logistic dose–response curves with 10 concentration points, each representing the average of two independent experiments.

2.8.2. Immunofluorescence Analysis

HeLa cells were fixed in 4% paraformaldehyde and subsequently permeabilized with 0.2% Triton X-100. Blocking was performed with 1% bovine serum albumin (BSA) for 10 min at room temperature. Cells were then incubated for 1 h at RT with primary antibodies targeting: γH2AX Ser139 (1:200, ab2893, Abcam, Cambridge, UK) or pRPA32 S4/8 (1:200, A300-245A, Bethyl Laboratories, Montgomery, TX, USA). Corresponding secondary antibody conjugated with AlexaFluor 647 donkey anti-rabbit IgG (H + L) was applied. Samples were mounted using ProLong™ Gold Antifade Mountant with DAPI (Thermo Fisher Scientific). High-resolution immunofluorescence imaging was performed on a Zeiss LSM 900 Airyscan2 confocal microscope equipped with a Plan-Apochromat 63×/1.4 NA oil immersion objective. Image acquisition and processing utilized Zen 3.9 software Zeiss, Jena, Germany). Quantification of fluorescent foci per nucleus was conducted using Fiji (ImageJ) software, version 2.14.0/1.54f.

2.8.3. Cell Viability Assays

HeLa, MCF-7, and MDA-MB-231 cells were seeded in triplicate into 96-well plates at densities of 1500, 3500, and 3000 cells per well, respectively, and allowed to adhere for 24 h. After 72 h of incubation, cells were fixed with 10% (*v*/*v*) trichloroacetic acid (TCA) and subsequently stained with 0.4% (*w*/*v*) sulforhodamine B (SRB) dissolved in 1% (*v*/*v*) acetic acid. Cell viability was quantified and expressed as a percentage relative to untreated control cells, which were normalized to 100%. Data represent the mean ± standard deviation (SD) from three independent experiments (n = 3). Statistical significance was

 https://doi.org/10.3390/biom16010012

determined using one-way ANOVA or Student's *t*-test, followed by appropriate multiple-comparison post hoc tests to evaluate differences between experimental groups.

2.9. Molecular Docking Studies

Molecular docking was performed with GOLD 5.1 using the GoldScore fitness function and the X-ray structure of CDK9 in complex with a co-crystallized ligand (PDB ID: 6GZH [23]). The binding pocket was defined as the region encompassing all residues located within 15 Å from the co-crystallized ligand. While the option for early termination of runs was disabled, ring corner flipping was permitted, and 100 genetic algorithm runs were executed. All other parameters were instead kept at the default configuration [24]. The obtained docking solutions were clustered using an RMSD threshold of 2.0 Å. Subsequently, clusters representing at least 5% of the poses were considered, and their representative conformations extracted. Among them, only poses forming at least one hydrogen bond with the hinge region of the protein were retained for further investigations.

2.10. Molecular Dynamics Simulation

MD simulation was carried out using AMBER version 24 [25] employing an NVIDIA RTX 3080 GPU. The protein was modeled with the ff14SB force field, while ligands were parameterized using GAFF, and their partial charges assigned through the AM1-BCC method with the Antechamber module. The protein–ligand complex was solvated in a rectangular parallelepiped water box with a 20.0 Å water cap using the TIP3P explicit solvent model. The system was then neutralized by adding Cl^- counterions. Prior to MD, a two-step minimization was performed. In the first stage, a 100 kcal/(mol·Å^2) position restraint was imposed on the whole complex, allowing only water molecules to relax through 5000 steps of steepest descent followed by conjugate gradient minimization. In the second stage, the entire system was minimized applying a harmonic restraint of 10 kcal/(mol·Å^2) only to the protein Cα atoms. The simulations were conducted using particle mesh Ewald (PME) electrostatics and periodic boundary conditions. Upon minimization, the heating MD step (0.5 ns) was performed using constraint-volume periodic boundary conditions with temperature rising from 0 to 300 K. Subsequently, 3 ns of equilibration were executed through constant pressure periodic boundary MD. The Langevin thermostat was employed in this step to keep the temperature of the system at the constant value of 300 K. Lastly, a production MD run of 196.5 ns was carried out under constant pressure and temperature conditions. In total, 200 ns of MD simulation were therefore performed, and the resulting trajectory was analyzed through the cpptraj program implemented in AMBER24. During the entire MD simulation, a harmonic force constant of 10 kcal/(mol·Å^2) was applied to restrain all Cα of the protein. This approach preserved the overall experimental conformation of the protein while still allowing for full side-chain flexibility and ligand adaptation within the binding site, thereby enabling an accurate evaluation of the stability of the ligand–protein interactions.

3. Results

3.1. Machine Learning Model Generation, Optimization, and Assessment

In order to develop ML models for identifying novel ligands potentially active against CDK9, we constructed a curated dataset by retrieving inhibitory activity data from ChEMBL31 [13], considering both CDK9 as an isolated target and as part of the biologically relevant CDK9/cyclin T1 complex. Only activity values reported with exact quantitative measures were retained. As IC_{50} data represented the majority of available measurements (~80%), they were prioritized for training, while the remaining data types were used for model evaluation. After being subjected to the data curation process, com-

https://doi.org/10.3390/biom16010012

pounds were classified as active and inactive according to two different classification settings, i.e., Lead-Oriented (LO) and Potency-Oriented (PO), and for each scheme, compounds with potency values falling within the corresponding classification thresholds were excluded (see Section 2). This decision aimed to avoid ambiguity near the classification boundary, where small fluctuations in measured activity could result in a switched class assignment. The two classification approaches resulted in two parallel workflows, which were carried out simultaneously, enabling a comparative analysis of model performance under different activity definitions. Due to an initial imbalance favoring the active class observed across all datasets, a two-step enrichment strategy was applied to increase the number of inactive compounds. In the first step, additional inactive molecules retrieved from ChEMBL were included, as detailed in the Section 2. Despite this integration, the datasets remained skewed, prompting the inclusion of an additional set of presumed inactive decoy compounds retrieved from the ZINC20 database [16] to further balance the class distribution. These decoys were selected to resemble drug-like molecules while being presumed inactive against CDK9. All added compounds underwent the same preprocessing protocol, and dataset integrity was ensured by eliminating any overlap between training and test sets. The resulting four datasets, two per classification scheme, were fully balanced in terms of classes, as summarized in Table 1.

Table 1. Composition of training and test sets for each classification scheme: Lead-Oriented (LO) and Potency-Oriented (PO).

	Lead-Oriented (LO)		
	Active	**Inactive**	**Total**
Training	1375	1375	2750
Test	279	279	558
	Potency-Oriented (PO)		
	Active	**Inactive**	**Total**
Training	447	447	894
Test	106	106	212

The generated models were obtained by representing the training set compounds through seven distinct molecular representations, including three types of molecular fingerprints (i.e., RDKit, Morgan, and PubChem fingerprint), and a set of 208 numerical descriptors calculated with RDKit [17]. To leverage complementary information from both structural patterns and physicochemical properties, we also constructed three hybrid representations by concatenating each fingerprint type with RDKit descriptors. These encodings were then used as input for five different ML algorithms: RF, SVM, KNN, GP, and MLP. By systematically combining all representations with all algorithms, a total of 70 models were generated, 35 for each activity threshold scheme. Every model combination was subjected to a dedicated hyperparameter optimization process aimed at maximizing predictive performance. Subsequently, each optimized model underwent a 5-fold CV procedure to assess its robustness across different data partitions. Results of the CV analysis are reported for each model in Tables S1–S5, using MCC as a performance metric. Overall, the CV performance indicates high stability across different data splits, as reflected by the low standard deviation values observed for most models. Interestingly, both the most complex (MLP, Table S5) and the simplest (KNN, Table S4) algorithms showed overall slightly lower performance compared to more traditional ML methods such as RF, SVM, and GP (Tables S1–S3). This trend may be partially explained by the nature of the dataset, which, although of moderate size, may not provide sufficient statistical power for deep learning architectures like MLP to generalize effectively. Deep learning models typically

 https://doi.org/10.3390/biom16010012

require large amounts of data to achieve optimal performance, while with datasets limited in size, they are prone to overfitting and instability. This is also reflected by the observed standard deviation for MLP-based models (Table S5), which appears to be consistently higher than the one observed for models based on other architectures (Tables S1–S4). On the other hand, KNN may struggle with the complexity and noise of chemical space due to its reliance on local similarity measures, which are more sensitive to class overlap and uneven data distributions. These limitations make intermediate-complexity models, such as RF, SVM, and GPC, better suited to effectively capture informative patterns in the datasets used in this study.

From the molecular representation perspective, hybrid encodings generally performed comparably to standard fingerprints, as particularly evident in the case of KNN models (Table S4). In certain instances, the addition of physicochemical descriptors to fingerprints slightly improved model performance. For example, the SVM model using HybridRR achieved a marginal improvement (ΔMCC = +0.07) over the corresponding model based solely on RDKit fingerprint (Table S2). However, this trend was not consistent across all model families: in some cases, especially when using the Morgan fingerprint, the integration of descriptors led to a reduction in performance, suggesting that the additional information may introduce noise or redundancy rather than a useful signal, depending on the algorithm and representation. As an example, PO models employing the RF and SVM algorithms (Tables S1 and S2) exhibited a decrease in performance, with ΔMCC values of -0.02 and -0.05, respectively. Regarding the data labeling strategy, models developed under the LO approach generally achieved higher MCC values across different algorithms, with the effect being more pronounced for MLP, KNN, and GP models (Tables S3–S5), and less marked for SVM and RF (Tables S1 and S2).

Upon CV analysis, each ML model was derived using the optimal hyperparameter setting previously identified and trained on the corresponding training set. To evaluate the generalization capabilities of each model on unseen data, thereby simulating the real scenario where models are required to predict the activity of unknown compounds, the threshold-specific test sets were used. Models were subsequently ranked based on the MCC values obtained during this step (Tables S6 and S7), and the best-performing one for each classification scheme was identified. For the PO approach, the highest MCC was achieved by the GP-RDKit fingerprint model (MCC = 0.71, Table S6), while for the LO scheme, the RF-Morgan fingerprint model yielded the best performance with an MCC of 0.61 (reported in Table S7). For the LO scheme, the Morgan fingerprint representation proved particularly effective, characterizing most top-ranking models, as reported in Table S7. Conversely, models relying on RDKit descriptors generally showed lower performance. Consistent with CV results, models based on the simplest and the most complex algorithms (KNN and MLP, respectively) performed generally worse than the others. A similar trend was observed for the PO models (Table S6), where most KNN- and MLP-based models were ranked among the lowest-performing ones, often combined with RDKit descriptors. Interestingly, the GP-RDKit fingerprint model of the PO classification scheme achieved the overall highest MCC (Table S6), reflecting the best capability to correctly predict both classes, and was therefore selected to guide the VS campaign.

3.2. Virtual Screening

Building on the results of the previous analysis, a VS study was carried out to identify novel potential inhibitors of the CDK9 protein. For this purpose, the best-performing model (GP-RDKit fingerprint) was employed. The screening was conducted on compounds from the AMS library, which was curated following the same procedure applied to the training and test sets. After curation, the over 7 million compounds obtained were represented as

https://doi.org/10.3390/biom16010012

RDKit fingerprints and submitted to the selected model to identify potential hit compounds. To guide the selection of the most promising candidates, all compounds predicted as active with a PS above 0.50 (411,881 molecules) were retrieved and ranked accordingly. To define an appropriate PS threshold for compound selection, the model performance on the test set was examined across values ranging from 0.50 to 0.90, focusing on how precision and recall varied (Figure 1).

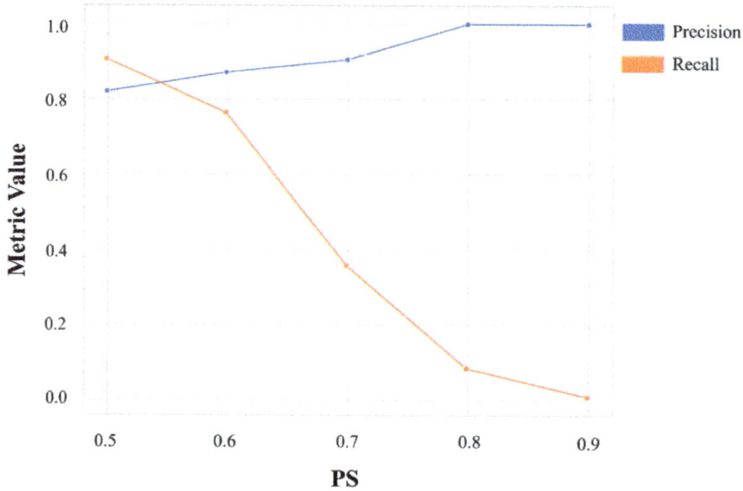

Figure 1. Precision and recall values achieved at different probability scores (PS) thresholds. Precision is reported in blue, while recall is reported in orange.

At a threshold of 0.8, the model achieved perfect precision, indicating a high confidence in the predicted actives and no false positive predictions. Based on this outcome, the 93 VS hits with PS $\geq$ 0.8 were prioritized for further filtering. To identify a representative subset of compounds capturing a broad chemical diversity within the 93 predicted hits, hierarchical agglomerative clustering was applied. Compounds were represented using RDKit fingerprint, and pairwise similarity was computed using the Tanimoto coefficient, a standard metric in cheminformatics for comparing molecular fingerprints (see Section 2). As a result, 21 clusters were identified by applying a distance cutoff of 0.30, a threshold informed by a previously reported protocol [10] and verified through visual inspection to be effective in the present scenario. From each cluster, the most promising compound was selected based on the associated PS and its commercial availability at the time of purchase. This filtering process led to the selection of 14 structurally diverse compounds, which were then purchased and submitted to enzymatic assays for experimental validation.

3.3. Enzymatic Assays

The 14 compounds selected from the VS campaign were subjected to enzymatic assays to evaluate their inhibitory activity against CDK9. The outcomes of this experimental validation are summarized in Table 2.

https://doi.org/10.3390/biom16010012

Table 2. Structure and experimental CDK9 inhibitory activity of the tested compounds. Roscovitine was used as a reference inhibitor.

Cpds.	Structure	Experimental IC_{50} (nM)
1	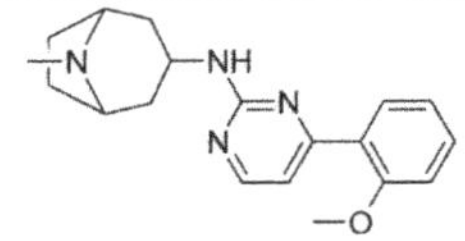	3510
2	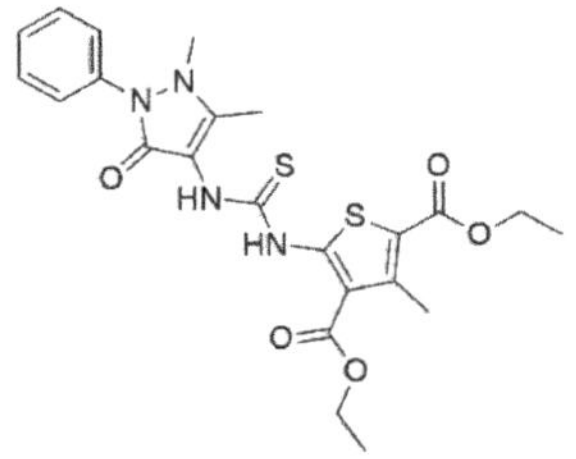	16,800
3	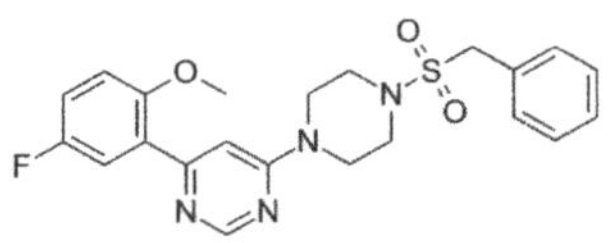	28,600
4	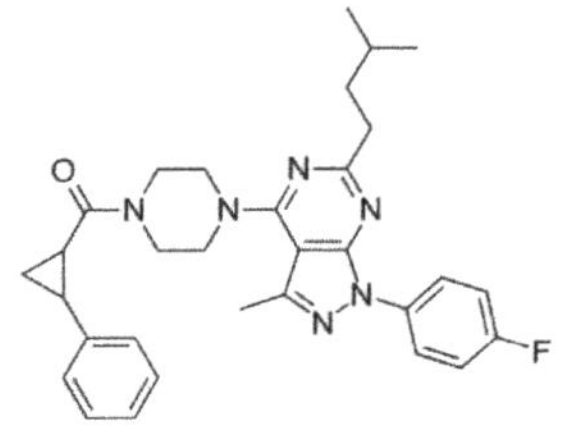	67,200
5	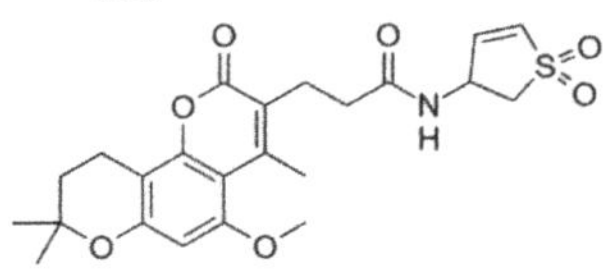	107,000
6	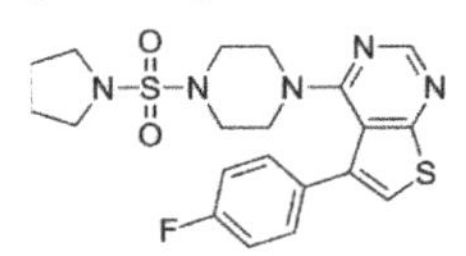	133,000
7	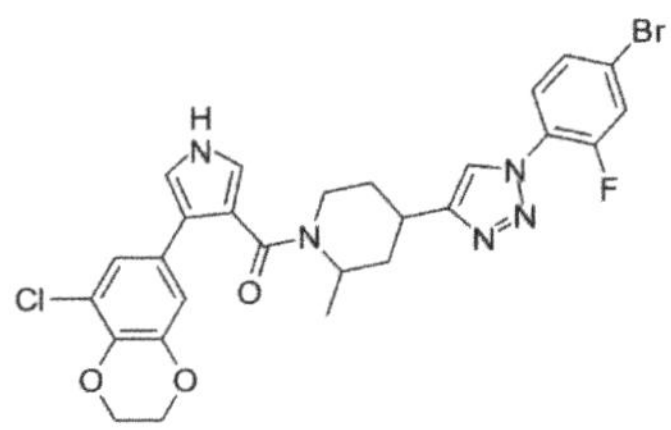	250,000
8	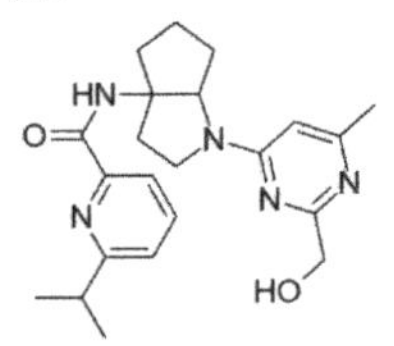	250,000

 https://doi.org/10.3390/biom16010012

Table 2. *Cont.*

Cpds.	Structure	Experimental IC_{50} (nM)
9	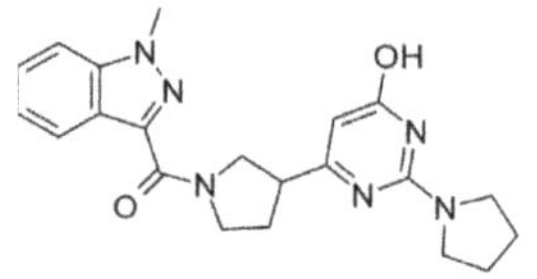	250,000
10	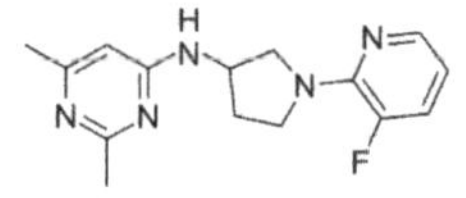	250,000
11	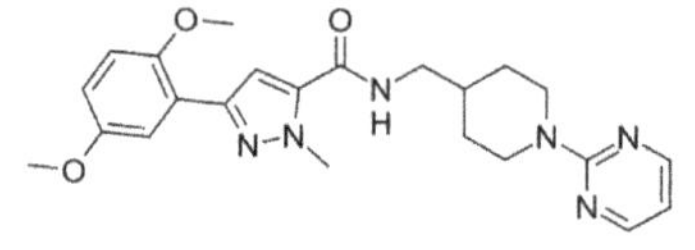	250,000
12	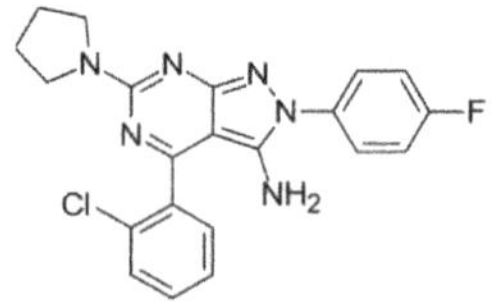	250,000
13	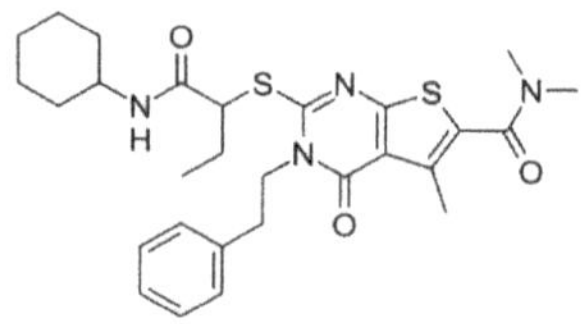	250,000
14	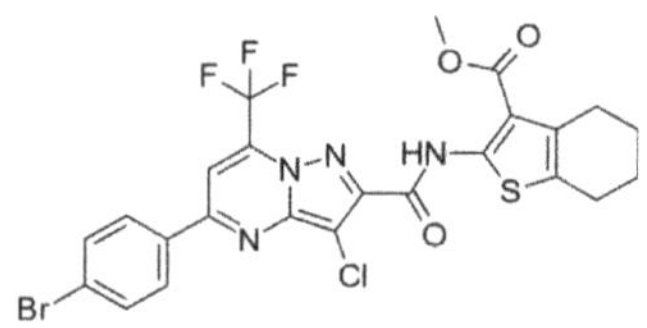	250,000
Roscovitine	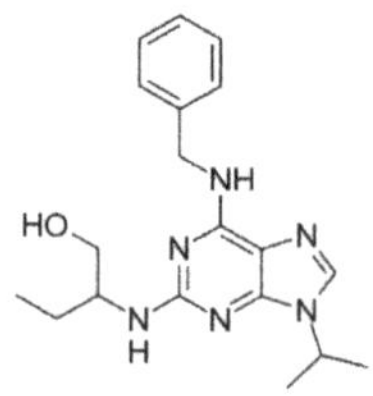	476

Around 30% of the tested compounds (4 out of 14) showed inhibitory activity below 100 μM, with compounds **1**, **2**, and **3** showing IC_{50} values below 50 μM. Most notably, two promising CDK9 inhibitors were identified, namely compounds **1** and **2**, which demonstrated inhibitory activity in the low micromolar range (with IC_{50} values of 3.51 μM and 16.80 μM, respectively). Consequently, these two compounds were prioritized for further investigation. To better assess their inhibitory effects in a more physiologically relevant context, both candidates were advanced to cellular assays.

3.4. Antiproliferative Assays

To evaluate the potential anticancer activity of compounds **1** and **2**, we initially performed cell viability assays on a panel of representative tumor cell lines, including HeLa,

https://doi.org/10.3390/biom16010012

MCF-7, and MDA-MB-231, which express different molecular characteristics [26]. Treatment with increasing concentrations of the two inhibitors revealed a clear dose-dependent reduction in cell viability. Specifically, compound **1** consistently displayed stronger antiproliferative activity across all tested cell lines (Figure 2A). In contrast, as reported in Figure 2B, compound **2** was less effective at comparable concentrations; however, when administered at higher doses, it demonstrated a significant ability to decrease cell viability, although with lower efficacy than compound **1**.

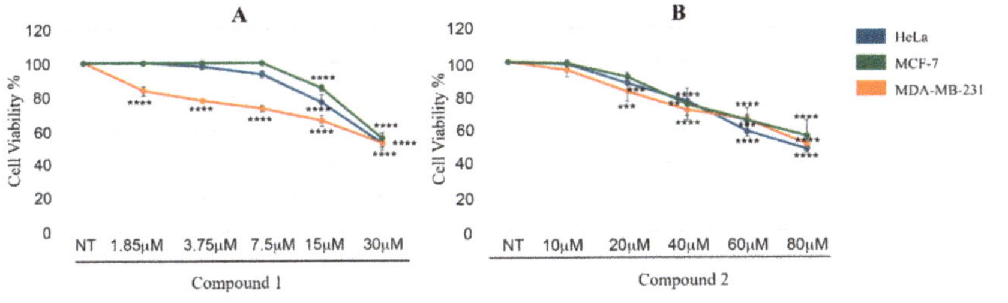

Figure 2. Antiproliferative effect measured as cell viability percentage achieved at different concentrations of compound **1** (**A**) and **2** (**B**) on the three selected cell lines. Results are presented as the mean ± standard deviation (SD) from three independent experiments. Statistically significant differences are indicated as ** $p < 0.01$, *** $p < 0.001$ and **** $p < 0.0001$.

These findings are consistent with the in vitro enzymatic assays, in which compound **1** displayed approximately a five-fold higher potency than compound **2**. Taken together, the obtained results indicate that compound **1** possesses not only stronger inhibitory activity but also greater therapeutic potential. Therefore, compound **1** was prioritized as the most promising candidate for further studies, with particular attention to its potential use in combination with conventional chemotherapy. Accordingly, we investigated whether CDK9 inhibition by compound **1** could enhance the activity of established anticancer agents. In particular, we focused on Camptothecin (CPT), a topoisomerase I inhibitor widely used in cancer therapy [27]. To explore its potential interaction with compound **1**, we first tested an intermediate concentration of the inhibitor (7.5 μM) in combination with a range of CPT doses (1.25–20 nM). This analysis (Figure 3A–C) provided an overview of the interaction profile and guided the selection of optimal conditions for subsequent experiments. On this basis, a fixed dose of CPT (5 nM) was chosen as a reference, as it produced a detectable cytotoxic effect while still allowing the evaluation of any additive or synergistic contribution from compound **1**. Under these conditions, we next tested an extended range of compound **1** concentrations (1.85–30 μM), which enabled a more detailed characterization of the combined treatment effects (Figure 3D–F).

As shown in Figure 3A–C, compound **1** consistently potentiated the cytotoxic activity of CPT across all tested concentrations relative to the DMSO control. On the other hand, as illustrated in Figure 3D–F, the use of a fixed CPT dose further amplified the cytotoxic effect of compound **1**, demonstrating the presence of an additive interaction between the two agents in all examined cell lines.

To further confirm that the cytotoxic effects of compound **1** are mediated through CDK9 inhibition, we performed cell viability assays in HeLa cells in which CDK9 expression was selectively reduced by siRNA-mediated knockdown, using non-targeting siRNA (siCTR) as a control. Cells with reduced CDK9 expression displayed higher survival upon treatment with compound **1** compared to control cells (results reported in Figure S2). This finding suggests that the presence of CDK9 contributes to the cytotoxic response induced by compound **1**. Consistently, when CDK9 is depleted, the overall impact of compound **1** is reduced. These results support the conclusion that the activity of compound **1** relies on

 https://doi.org/10.3390/biom16010012

CDK9 inhibition, thereby strengthening the evidence for CDK9 as the primary mediator of the herein identified inhibitory effect.

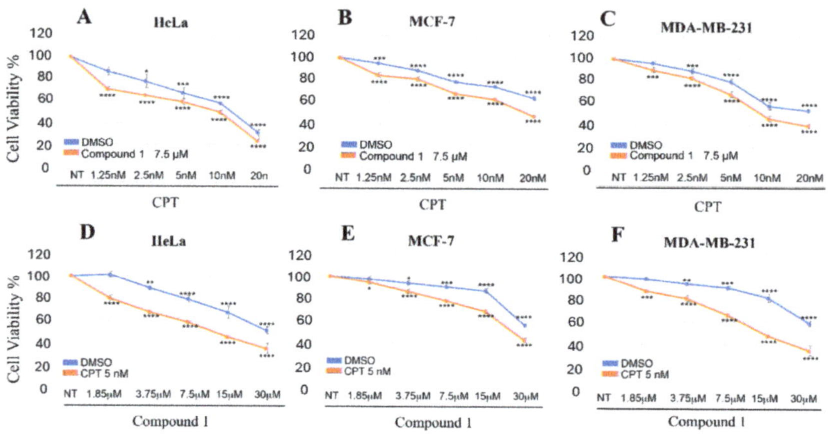

Figure 3. Cell viability assay of HeLa, MCF-7, and MDA-MB-231 cells treated with indicated doses of CPT and compound **1**, alone or in combination, for 72 h. * $p < 0.5$, ** $p < 0.01$, *** $p < 0.001$ and **** $p < 0.0001$. Results are presented as the mean ± SD from three independent experiments: (**A–C**) report the results obtained with the first experimental setting, where the concentration of compound **1** was held fixed; (**D–F**) refer to the results obtained with the second setting, in which CPT was kept at a fixed concentration.

To gain deeper insight into the mechanism underlying the combined effect of compound **1** and CPT, we examined whether the former could influence the extent of DNA damage induced by CPT. Immunofluorescence analysis of γ-H2AX foci, a well-established marker of double-strand breaks [28], revealed a stronger DNA damage response in cells treated with the drug combination compared to CPT alone, as reported in Figure S3. These findings suggest that compound **1** potentiates the genotoxic effect of CPT. Furthermore, as CPT primarily induces double-strand breaks during the S phase, which are normally repaired through homologous recombination (HR), we next asked whether compound **1** might interfere with this repair pathway. To address this, HeLa cells were pretreated with compound **1** or DMSO for one hour, followed by a two-hour exposure to CPT, and the phosphorylation of RPA32 at S4/8, a marker of HR activity [29], was assessed by immunofluorescence. As shown in Figure S4, no significant changes in pRPA32 signal were observed upon treatment with compound **1**, indicating that HR processivity remained unaltered. Collectively, these results indicate that compound **1** potentiates CPT-induced cytotoxicity by enhancing DNA damage, without impairing HR repair.

3.5. Molecular Modeling Studies

To propose a potential binding mode of compound **1** within the CDK9 active site, we carried out a molecular docking study followed by MD simulation using the X-ray structure of CDK9 (PDB ID: 6GZH [23]). Initially, compound **1** was subjected to docking analysis using GOLD software version 5.1 with the GoldScore fitness function. Clustering of the resulting docking poses with a root mean squared deviation (RMSD) cutoff of 2.0 Å (see Section 2) generated only one reliable binding model for the ligand, characterized by a highly predominant population (97%), and showing at least one hydrogen bond within the hinge region of the protein. This pose was then subjected to 200 ns of MD simulation to assess the stability of the predicted binding conformation and to examine the key interactions with CDK9. Throughout the simulation, the ligand maintained a stable

https://doi.org/10.3390/biom16010012

orientation, with an average RMSD relative to the initial docking pose of approximately 1 Å (Figure 4).

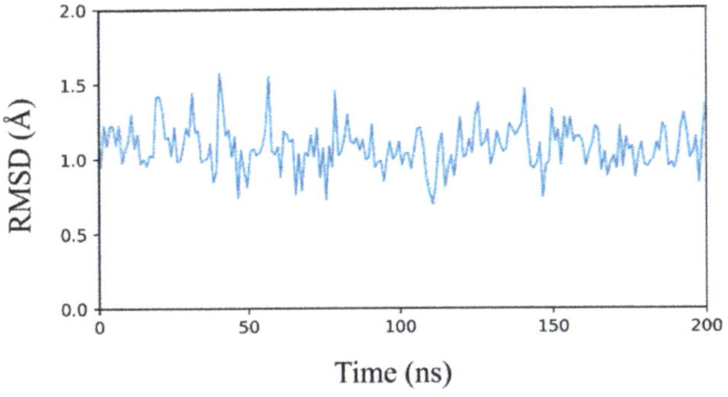

Figure 4. Analysis of MD simulation for compound **1**.

The remarkable stability of the ligand disposition, as well as its interactions with the protein binding site, throughout the whole simulation, confirmed the reliability of the predicted binding mode of compound **1** into CDK9. Figure 5 shows the minimized average structure of CDK9 bound to compound **1** obtained from the last 100 ns of MD simulation, whereas Figure S5 provides a molecular surface representation of the binding site along with compound **1**, highlighting the shape complementarity between the ligand and the surrounding pocket. The predicted binding mode of the ligand is mainly stabilized by three hydrogen bonds involving two residues of the protein hinge region. In particular, residue C93 is involved in two distinct hydrogen bonding interactions. The first occurs between its backbone nitrogen and the pyrimidine nitrogen of the ligand, while the second involves the backbone oxygen of the residue and the NH linker connecting the tropane and pyrimidine moieties of the inhibitor. Additionally, the tropane core further stabilizes the ligand binding through the formation of a third hydrogen bond with the backbone oxygen of residue E94. Beyond these polar interactions, the tropane ring engages in hydrophobic contacts mainly with residue I18, whereas the pyrimidine ring establishes hydrophobic interactions with A39 and L143. Furthermore, the methoxyphenyl ring of the ligand is accommodated within a hydrophobic pocket defined by residues F23, V26, and F90, forming notable hydrophobic interactions with V26 and π–π stacking interactions with F23 and F90, collectively contributing to the stabilization of the ligand–protein complex.

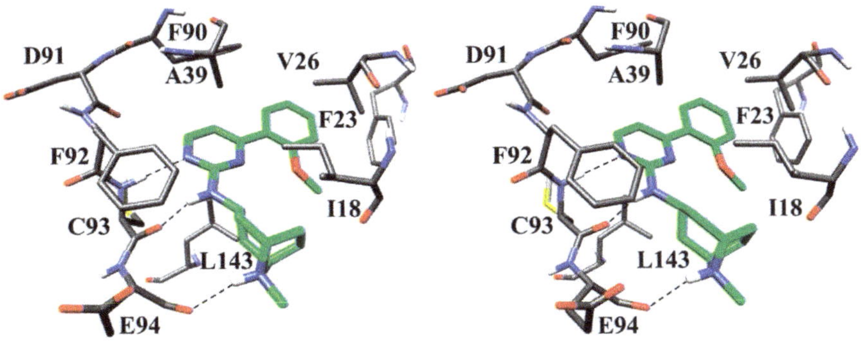

Figure 5. Stereoscopic representation of the minimized average structure of compound **1** in complex with CDK9. Hydrogen bonds are represented as black dashed lines. Protein residues surrounding the ligand, which is represented in green, are shown as grey sticks.

https://doi.org/10.3390/biom16010012

4. Discussion

CDK9 plays a central role in transcriptional elongation and DNA repair and exerts its biological activity primarily in complex with cyclin T1, forming the catalytic core of P-TEFb. Beyond its well-established role in cancer, CDK9 has also been implicated in several other pathological conditions characterized by transcriptional dysregulation, including inflammatory, autoimmune, and cardiovascular diseases, underscoring its relevance as a multifaceted therapeutic target across diverse biomedical areas.

Driven by the objective of exploiting the power of ML to identify novel and promising compounds with high therapeutic potential as CDK9 inhibitors, we first collected and curated bioactivity data for both CDK9 alone and the CDK9/cyclin T1 complex. This strategy ensured that the models were both trained on biologically meaningful information and built upon a sufficiently large dataset, two key requirements for obtaining robust and reliable ML models. Upon data standardization, two different classification schemes (LO and PO, as previously detailed) were defined with the aim of exploring model performance under varying levels of pharmacological stringency. The LO scheme was adapted from a previous study conducted on a closely related kinase target [11], where these cutoffs proved effective. Furthermore, in VS contexts, micromolar or sub-micromolar potency is generally associated with compounds of interest for further optimization [30]. However, considering the distribution of the collected bioactivity data (Figure S1), which revealed that most compounds tested against CDK9 exhibited strong potency, a second and more selective classification scheme (PO) was also implemented. The purpose of setting higher potency thresholds with a broader window between them was to train an alternative family of models using highly potent compounds, thereby focusing the learning process on distinguishing molecules with strong inhibitory potential from weaker or inactive ones.

The derived datasets were embedded in seven different molecular representations to evaluate how diverse structural encodings could influence model performance. In addition to standard molecular fingerprints and numerical descriptors, hybrid representations combining them were also generated. This design aimed to explore whether integrating the binary substructure information captured by fingerprints with the continuous physicochemical properties described by molecular descriptors could provide a more informative basis for training predictive ML models. Nevertheless, the results obtained showed that the combination of the two representations led to slight improvements only for certain models, while in others, it likely introduced noise rather than useful information, resulting in a decrease in predictive performance compared with non-hybrid representations. In a broader perspective, this observation underscores the importance of developing multiple models based on different algorithms and molecular encodings, since it is not possible to determine a priori which combination will yield the best performance for a given predictive task. Subsequently, considering the importance of evaluating models in terms of their applicability and generalization to novel compounds, all developed models were further validated using the corresponding test set. The best-performing model was identified as the GP-RDKit fingerprint model developed under the PO approach. In addition to its superior performance, this model was trained using a stricter activity threshold, potentially enhancing its ability to prioritize highly promising candidates. Collectively, these aspects supported its selection as the reference model for the in silico screening campaign. Additionally, the chemical space used to train and validate this model was characterized by projecting RDKit fingerprints for the PO dataset onto the first two principal components obtained from the PCA analysis. This projection (Figure S6) displays a broad dispersion of structures, reflecting substantial structural variability and a high degree of chemical diversity within the dataset.

https://doi.org/10.3390/biom16010012

The 14 most promising candidates resulting from the VS were purchased and subjected to enzymatic assays to experimentally validate the ML-based workflow. The obtained results confirmed the effectiveness of the proposed approach, as two candidates, namely compounds **1** and **2**, displayed low-micromolar inhibitory potency toward CDK9, a level of activity generally regarded as highly promising in early stages of drug discovery. These findings further illustrate the value of a rapid, data-driven, and cost-efficient computational strategy capable of identifying biologically active molecules without the need for extensive experimental screening, in accordance with previous evidence supporting the effectiveness of such approaches.

Considering the promising inhibitory activity of these two compounds, their potential therapeutic relevance was subsequently investigated through biological assays conducted in cellular models. Compound **1**, in particular, emerged as the most promising candidate, showing consistent antiproliferative effects across multiple cancer cell lines. In light of the growing clinical evidence that rational drug combinations can enhance therapeutic efficacy and help overcome resistance mechanisms, we next evaluated its interaction with CPT, a chemotherapeutic agent commonly employed in cancer treatment. The results revealed an additive effect between CDK9 inhibition by compound **1** and CPT treatment, suggesting that the two agents act cooperatively to increase cytotoxicity. Notably, the magnitude of this response varied among the tested cell lines, indicating that the characteristic genetic and molecular backgrounds of each cell type could influence their combined activity. Such variability is particularly relevant in the context of precision oncology and personalized therapy, where the design and selection of cancer treatments increasingly depend on the patient's specific genetic profile. Finally, additional experimental assays were performed to better characterize the mechanism underlying this combined effect, showing that the cytotoxic activity of compound **1** is dependent on CDK9 inhibition and that its co-administration with CPT amplifies CPT-induced DNA damage without compromising the efficiency of HR repair.

As a final step, a computational investigation combining molecular docking and MD simulation was carried out to provide a structural rationale for the observed biological activity. The proposed binding mode, deemed reliable in light of its remarkable stability, revealed that compound **1** forms a consistent network of hydrogen bonds and hydrophobic interactions with key residues within the CDK9 binding pocket.

Within this framework, previously reported computational efforts offer useful points of comparison. Lately, several computational strategies have been successfully applied to identify novel CDK9 inhibitors [31–35] and to characterize the activity profiles of small molecules against diverse kinase targets [12,36]. Moreover, ML-based VS has repeatedly demonstrated its ability to efficiently prioritize promising compounds with reduced experimental cost and time across multiple target classes, including kinases [10,11,37]. However, despite these advances, the specific case of CDK9 still presents unmet needs. Existing studies contribute valuable data and identify new molecules without, however, providing accessible predictive tools, ultimately limiting the practical adoption of ML-based methods by the wide scientific community for new discovery efforts. To the best of our knowledge, only one study provides a directly comparable baseline, as it publicly releases an ML classifier model for predicting CDK9 activity [38]. When evaluated on our test set, this model reached an MCC of 0.47, whereas the classifier developed under the PO strategy (GP–RDKit fingerprints), which we used to guide the VS campaign and was made publicly available, achieved an MCC of 0.71. Viewed within the broader scientific landscape, the relevance of the herein presented work becomes evident, as it addresses a persistent gap in a field where accelerating the discovery of new CDK9 inhibitors remains a critical priority. Furthermore, when considered together with the structural insight provided by our compu-

 https://doi.org/10.3390/biom16010012

tational analysis, which may support future optimization efforts, the overall contribution of this study is further enhanced, underscoring its value in advancing computationally guided strategies for the identification and refinement of novel CDK9 inhibitors.

5. Conclusions

CDK9 is a multifunctional kinase regulating key processes from DNA repair to transcriptional elongation. By sustaining the expression of oncogenes and short-lived anti-apoptotic proteins, CDK9 supports cancer cell survival, proliferation, and resistance to conventional therapies. Its overexpression in multiple solid and hematological malignancies is indeed associated with poor prognosis and increased metastatic potential. Moreover, CDK9 plays a role beyond oncology, being implicated in other pathological conditions characterized by transcriptional dysregulation. Therefore, CDK9 represents a highly relevant therapeutic target, as inhibiting its activity can disrupt critical survival pathways and holds promise for the development of novel therapeutic approaches in diseases where transcriptional control is altered.

In this study, a systematic process of data collection, preprocessing, and model optimization was applied to generate a total of seventy ML models by combining different algorithms, molecular representations, and classification thresholds. The performance of all models was evaluated to identify the best predictor, which was then used to conduct a VS campaign aimed at discovering novel CDK9 inhibitors. Enzymatic assays performed on the most promising candidates resulted from the VS revealed that two compounds, namely compounds **1** and **2**, exhibited inhibitory activity in the low micromolar range, with IC_{50} values of 3.51 μM and 16.80 μM, respectively. Considering that inhibitory activity in the low micromolar range is regarded as both encouraging and highly promising at this stage of drug discovery, these two compounds were advanced to cell-based assays on three cancer cell lines, HeLa, MCF-7, and MDA-MB-231, selected to represent different tumor types and genetic contexts. Among them, compound **1** displayed significant cytotoxic activity. Further experiments investigated the effect of compound **1** in combination with CPT. The combined treatment enhanced cytotoxicity compared to single-agent treatments, suggesting a cooperative mechanism that also amplifies DNA damage. Notably, the variable response observed among the tested cell lines indicates that the efficacy of the combination may depend on genetic factors, a concept aligned with the principles of precision oncology and personalized therapy. Finally, structure-based analyses were carried out to predict a plausible binding mode for compound **1**, providing valuable insights that may serve as a foundation for future optimization efforts aimed at developing novel, highly potent CDK9 inhibitors. To encourage transparency and support future studies, the final predictive model was made openly available on GitHub (https://github.com/MMVSL/CDK9Screen, accessed on 25 October 2025), providing researchers with an accessible tool to evaluate the potential inhibitory activity of novel CDK9 candidates.

Supplementary Materials: The following supporting information can be downloaded at: https://www.mdpi.com/article/10.3390/biom16010012/s1, Figure S1: Distribution of activity values for training set compounds; Table S1: Cross validation results for RF-based models; Table S2: Cross validation results for SVM-based models; Table S3: Cross validation results for GP-based models; Table S4: Cross validation results for KNN-based models; Table S5: Cross validation results for MLP-based models; Table S6: Test set performance of PO models; Table S7: Test set performance of LO models; Figure S2: Cell viability assays in HeLa cells transfected with siCTR or siCDK9; Figure S3: Immunofluorescence analysis of γ-H2AX foci; Figure S4: Analysis of HR activity; Figure S5: Molecular surface representation of the CDK9 binding site in complex with compound 1; Figure S6: PCA projection of PO dataset.

https://doi.org/10.3390/biom16010012

Author Contributions: Conceptualization, A.G. and T.T.; methodology, L.P., C.P., G.P. and M.D.S.; validation, G.M.N., L.A. and A.M.; formal analysis, L.P., G.B., C.G., A.M. and G.P.; investigation, G.M.N., G.B. and M.D.S.; resources, A.G., L.A. and T.T.; data curation, L.P., C.P., M.D.S., G.B. and C.G.; writing—original draft preparation, L.P., C.P., M.D.S. and A.M.; writing—review and editing, L.P., G.P., G.B., C.G. and T.T.; visualization, C.G., G.M.N. and C.P.; supervision, A.G., T.T. and L.A.; project administration, T.T. and L.A.; funding acquisition, L.A. and T.T. All authors have read and agreed to the published version of the manuscript.

Funding: This research received funding from the Sbarro Health Research Organization, the Italian Ministry of Health via the Ricerca Corrente 2025 Grant 3/29_25 and the National Recovery and Resilience Plan (PNRR), Mission 4, Component 2, Investment 1.4 "National Centre for HPC, Big Data and Quantum Computing"–Spoke 7 "Materials and Molecular Sciences" (European Union–NextGenerationEU).

Institutional Review Board Statement: Not applicable.

Informed Consent Statement: Not applicable.

Data Availability Statement: Original microscopy images have been deposited at Zenodo at https://doi.org/10.5281/zenodo.17472637. The ML model developed and used for the identification of novel CDK9 inhibitors was made publicly available as a predictive tool on GitHub (https://github.com/MMVSL/CDK9Screen, accessed on 25 October 2025).

Conflicts of Interest: The authors declare no conflicts of interest.

References

1. Mandal, R.; Becker, S.; Strebhardt, K. Targeting CDK9 for Anti-Cancer Therapeutics. *Cancers* **2021**, *13*, 2181. [CrossRef]
2. Alfano, L.; Iannuzzi, C.A.; Barone, D.; Forte, I.M.; Ragosta, M.C.; Cuomo, M.; Mazzarotti, G.; Dell'Aquila, M.; Altieri, A.; Caporaso, A.; et al. CDK9-55 Guides the Anaphase-Promoting Complex/Cyclosome (APC/C) in Choosing the DNA Repair Pathway Choice. *Oncogene* **2024**, *43*, 1263–1273. [CrossRef] [PubMed]
3. Bacon, C.W.; D'Orso, I. CDK9: A Signaling Hub for Transcriptional Control. *Transcription* **2019**, *10*, 57–75. [CrossRef] [PubMed]
4. Egloff, S. CDK9 Keeps RNA Polymerase II on Track. *Cell. Mol. Life Sci.* **2021**, *78*, 5543–5567. [CrossRef]
5. Anshabo, A.T.; Milne, R.; Wang, S.; Albrecht, H. CDK9: A Comprehensive Review of Its Biology, and Its Role as a Potential Target for Anti-Cancer Agents. *Front. Oncol.* **2021**, *11*, 678559. [CrossRef]
6. Mo, C.; Wei, N.; Li, T.; Ahmed Bhat, M.; Mohammadi, M.; Kuang, C. CDK9 Inhibitors for the Treatment of Solid Tumors. *Biochem. Pharmacol.* **2024**, *229*, 116470. [CrossRef]
7. Kryštof, V.; Chamrád, I.; Jorda, R.; Kohoutek, J. Pharmacological Targeting of CDK9 in Cardiac Hypertrophy. *Med. Res. Rev.* **2010**, *30*, 646–666. [CrossRef]
8. Hellvard, A.; Zeitlmann, L.; Heiser, U.; Kehlen, A.; Niestroj, A.; Demuth, H.U.; Koziel, J.; Delaleu, N.; Potempa, J.; Mydel, P. Inhibition of CDK9 as a Therapeutic Strategy for Inflammatory Arthritis. *Sci. Rep.* **2016**, *6*, 31441. [CrossRef] [PubMed]
9. Toure, M.A.; Koehler, A.N. Addressing Transcriptional Dysregulation in Cancer through CDK9 Inhibition. *Biochemistry* **2023**, *62*, 1114–1123. [CrossRef]
10. Galati, S.; Di Stefano, M.; Bertini, S.; Granchi, C.; Giordano, A.; Gado, F.; Macchia, M.; Tuccinardi, T.; Poli, G. Identification of New GSK3β Inhibitors through a Consensus Machine Learning-Based Virtual Screening. *Int. J. Mol. Sci.* **2023**, *24*, 17233. [CrossRef]
11. Di Stefano, M.; Galati, S.; Ortore, G.; Caligiuri, I.; Rizzolio, F.; Ceni, C.; Bertini, S.; Bononi, G.; Granchi, C.; Macchia, M.; et al. Machine Learning-Based Virtual Screening for the Identification of Cdk5 Inhibitors. *Int. J. Mol. Sci.* **2022**, *23*, 10653. [CrossRef]
12. Di Stefano, M.; Piazza, L.; Poles, C.; Galati, S.; Granchi, C.; Giordano, A.; Campisi, L.; Macchia, M.; Poli, G.; Tuccinardi, T. KinasePred: A Computational Tool for Small-Molecule Kinase Target Prediction. *Int. J. Mol. Sci.* **2025**, *26*, 2157. [CrossRef]
13. Mendez, D.; Gaulton, A.; Bento, A.P.; Chambers, J.; De Veij, M.; Félix, E.; Magariños, M.P.; Mosquera, J.F.; Mutowo, P.; Nowotka, M.; et al. ChEMBL: Towards Direct Deposition of Bioassay Data. *Nucleic Acids Res.* **2019**, *47*, D930–D940. [CrossRef] [PubMed]
14. Galati, S.; Di Stefano, M.; Macchia, M.; Poli, G.; Tuccinardi, T. MolBook UNIPI–Create, Manage, Analyze, and Share Your Chemical Data for Free. *J. Chem. Inf. Model.* **2023**, *63*, 3977–3982. [CrossRef]
15. Galati, S.; Di Stefano, M.; Piazza, L.; Poles, C.; Macchia, M.; Tuccinardi, T. MolBook Pro 2025. Available online: https://molbookpro.farm.unipi.it/ (accessed on 10 January 2025).
16. Irwin, J.J.; Tang, K.G.; Young, J.; Dandarchuluun, C.; Wong, B.R.; Khurelbaatar, M.; Moroz, Y.S.; Mayfield, J.; Sayle, R.A. ZINC20-A Free Ultralarge-Scale Chemical Database for Ligand Discovery. *J. Chem. Inf. Model.* **2020**, *60*, 6065–6073. [CrossRef]
17. Landrum, G. RDKit: Open-Source Cheminformatics. Available online: https://www.rdkit.org (accessed on 1 February 2023).

18. Pedregosa, F.; Varoquaux, G.; Gramfort, A.; Michel, V.; Thirion, B.; Grisel, O.; Blondel, M.; Müller, A.; Nothman, J.; Louppe, G.; et al. Scikit-Learn: Machine Learning in Python. *J. Mach. Learn. Res.* **2011**, *12*, 2825–2830.
19. Ralaivola, L.; Swamidass, S.J.; Saigo, H.; Baldi, P. Graph Kernels for Chemical Informatics. *Neural Netw.* **2005**, *18*, 1093–1110. [CrossRef] [PubMed]
20. Matthews, B.W. Comparison of the Predicted and Observed Secondary Structure of T4 Phage Lysozyme. *Biochim. Biophys. Acta (BBA)-Protein Struct.* **1975**, *405*, 442–451. [CrossRef]
21. Bajusz, D.; Rácz, A.; Héberger, K. Why is Tanimoto Index an Appropriate Choice for Fingerprint-Based Similarity Calculations? *J. Cheminform.* **2015**, *7*, 20. [CrossRef]
22. Demšar, J.; Erjavec, A.; Hočevar, T.; Milutinovič, M.; Možina, M.; Toplak, M.; Umek, L.; Zbontar, J.; Zupan, B. Orange: Data Mining Toolbox in Python. *J. Mach. Learn. Res.* **2013**, *14*, 2349–2353.
23. Minzel, W.; Venkatachalam, A.; Fink, A.; Hung, E.; Brachya, G.; Burstain, I.; Shaham, M.; Rivlin, A.; Omer, I.; Zinger, A.; et al. Small Molecules Co-Targeting CKIα and the Transcriptional Kinases CDK7/9 Control AML in Preclinical Models. *Cell* **2018**, *175*, 171–185.e25. [CrossRef] [PubMed]
24. D'Ascenzio, M.; Secci, D.; Carradori, S.; Zara, S.; Guglielmi, P.; Cirilli, R.; Pierini, M.; Poli, G.; Tuccinardi, T.; Angeli, A.; et al. 1,3-Dipolar Cycloaddition, HPLC Enantioseparation, and Docking Studies of Saccharin/Isoxazole and Saccharin/Isoxazoline Derivatives as Selective Carbonic Anhydrase IX and XII Inhibitors. *J. Med. Chem.* **2020**, *63*, 2470–2488. [CrossRef]
25. Case, D.A.; Cheatham, T.E.; Darden, T.; Gohlke, H.; Luo, R.; Merz, K.M.; Onufriev, A.; Simmerling, C.; Wang, B.; Woods, R.J. The Amber Biomolecular Simulation Programs. *J. Comput. Chem.* **2005**, *26*, 1668–1688. [CrossRef] [PubMed]
26. Mota, A.d.L.; Evangelista, A.F.; Macedo, T.; Oliveira, R.; Scapulatempo-Neto, C.; Vieira, R.A.d.C.; Marques, M.M.C. Molecular Characterization of Breast Cancer Cell Lines by Clinical Immunohistochemical Markers. *Oncol. Lett.* **2017**, *13*, 4708–4712. [CrossRef]
27. Kamle, M.; Pandhi, S.; Mishra, S.; Barua, S.; Kurian, A.; Mahato, D.K.; Rasane, P.; Büsselberg, D.; Kumar, P.; Calina, D.; et al. Camptothecin and Its Derivatives: Advancements, Mechanisms and Clinical Potential in Cancer Therapy. *Med. Oncol.* **2024**, *41*, 263. [CrossRef]
28. Sharma, A.; Singh, K.; Almasan, A. Histone H2AX Phosphorylation: A Marker for DNA Damage. *Methods Mol. Biol.* **2012**, *920*, 613–626. [CrossRef]
29. Sartori, A.A.; Lukas, C.; Coates, J.; Mistrik, M.; Fu, S.; Bartek, J.; Baer, R.; Lukas, J.; Jackson, S.P. Human CtIP Promotes DNA End Resection. *Nature* **2007**, *450*, 509–514. [CrossRef] [PubMed]
30. Zhu, T.; Cao, S.; Su, P.C.; Patel, R.; Shah, D.; Chokshi, H.B.; Szukala, R.; Johnson, M.E.; Hevener, K.E. Hit Identification and Optimization in Virtual Screening: Practical Recommendations Based on a Critical Literature Analysis. *J. Med. Chem.* **2013**, *56*, 6560–6572. [CrossRef]
31. Yang, Q.; Shen, Z.; Chen, R.; Yao, C.; Pan, Z.; Wu, D.; Dong, X. Identification of Potent CDK9 Inhibitors with Novel Skeletons via Virtual Screening, Biological Evaluation, and Molecular Dynamics Simulation. *ACS Med. Chem. Lett.* **2024**, *15*, 1654–1661. [CrossRef]
32. Chen, G.; He, H.; Zhao, L.; Chen, K.-B.; Li, S.; Chen, C.Y.-C. Adaptive Boost Approach for Possible Leads of Triple-Negative Breast Cancer. *Chemom. Intell. Lab. Syst.* **2022**, *231*, 104690. [CrossRef]
33. Wu, M.; Han, J.; Liu, Z.; Zhang, Y.; Huang, C.; Li, J.; Li, Z. Identification of Novel CDK 9 Inhibitors Based on Virtual Screening, Molecular Dynamics Simulation, and Biological Evaluation. *Life Sci.* **2020**, *258*, 118228. [CrossRef]
34. Asamitsu, K.; Hirokawa, T.; Okamoto, T. Drug Design for Cyclin-Dependent Kinase 9 (CDK9) Inhibitors in Silico. *Biochem. Biophys. Rep.* **2025**, *42*, 101988. [CrossRef]
35. Singh, P.; Kumar, V.; Jung, T.S.; Lee, J.S.; Lee, K.W.; Hong, J.C. Uncovering Potential CDK9 Inhibitors from Natural Compound Databases through Docking-Based Virtual Screening and MD Simulations. *J. Mol. Model.* **2024**, *30*, 267. [CrossRef] [PubMed]
36. Liu, C.; Kutchukian, P.; Nguyen, N.D.; AlQuraishi, M.; Sorger, P.K. A Hybrid Structure-Based Machine Learning Approach for Predicting Kinase Inhibition by Small Molecules. *J. Chem. Inf. Model.* **2023**, *63*, 5457–5472. [CrossRef] [PubMed]
37. Piazza, L.; Di Stefano, M.; Poles, C.; Bononi, G.; Poli, G.; Renzi, G.; Galati, S.; Giordano, A.; Macchia, M.; Carta, F.; et al. A Machine Learning Platform for Isoform-Specific Identification and Profiling of Human Carbonic Anhydrase Inhibitors. *Pharmaceuticals* **2025**, *18*, 1007. [CrossRef]
38. Zhang, H.; Huang, J.; Chen, R.; Cai, H.; Chen, Y.; He, S.; Xu, J.; Zhang, J.; Wang, L. Ligand- and Structure-Based Identification of Novel CDK9 Inhibitors for the Potential Treatment of Leukemia. *Bioorg. Med. Chem.* **2022**, *72*, 116994. [CrossRef]

 https://doi.org/10.3390/biom16010012

biomolecules

MDPI

Article

Bile Acids Are Potential Negative Allosteric Modulators of M1 Muscarinic Receptors

Wenbo Yu [1,2,3,*], Alexander D. MacKerell, Jr. [1,2,3], David J. Weber [2,3,4,5] and Jean-Pierre Raufman [4,5,6,7,*]

1 Computer-Aided Drug Design Center, Department of Pharmaceutical Sciences, School of Pharmacy, University of Maryland Baltimore, Baltimore, MD 21201, USA; amackerell@rx.umaryland.edu
2 Institute for Bioscience and Biotechnology Research (IBBR), Rockville, MD 20850, USA; dweber@som.umaryland.edu
3 Center for Biomolecular Therapeutics (CBT), School of Medicine, University of Maryland Baltimore, Baltimore, MD 21201, USA
4 Department of Biochemistry and Molecular Biology, University of Maryland School of Medicine, Baltimore, MD 21201, USA
5 Marlene and Stewart Greenebaum Cancer Center, University of Maryland School of Medicine, Baltimore, MD 21201, USA
6 VA Maryland Healthcare System, Baltimore, MD 21201, USA
7 Department of Medicine, Division of Gastroenterology and Hepatology, University of Maryland School of Medicine, Baltimore, MD 21201, USA
* Correspondence: wyu@rx.umaryland.edu (W.Y.); jraufman@som.umaryland.edu (J.-P.R.)

Abstract: The proposed physiological roles of bile acids have expanded beyond the digestion of fats to encompass cell signaling via the activation of a variety of nuclear and plasma membrane receptors in multiple organ systems. The current in silico study was inspired by previous observations from our group and others that bile acids interact functionally with cardiac, pulmonary, and gastrointestinal muscarinic receptors and more recent work demonstrating allosteric binding of cholesterol, the parent molecule for bile acid synthesis, to M_1 muscarinic receptors (M_1R). Here, we computationally tested the hypothesis that bile acids can allosterically bind to M_1R and thereby modulate receptor activation. Utilizing de novo site identification by the ligand competitive saturation (SILCS) method, putative novel allosteric binding sites of bile acid targeting M_1R were identified. Molecular dynamics simulations were used to uncover the molecular details of the activation mechanism of M_1R due to agonist binding along with allosteric modulation of bile acids on M_1R activation. Allosteric binding of bile acids and their glycine and taurine conjugates to M_1R negatively impacts the activation process, findings consistent with recent reports that M_1R expression and activation inhibit colon cancer cell proliferation. Thus, bile acids may augment colon cancer risk by inhibiting the tumor suppressor actions of M_1R. When validated experimentally, these findings are anticipated to shed light on our understanding of how bile acids in the membrane microenvironment can allosterically modulate the function of M_1R and possibly other G protein-coupled receptors.

Keywords: bile acids; muscarinic receptors; cholinergic signaling; allosteric binding; site identification by ligand competitive saturation; docking; molecular dynamics

1. Introduction

Bile acids are amphipathic hydroxylated steroid byproducts of hepatic cholesterol metabolism that act as detergents which facilitate lipid absorption by emulsifying dietary fats. Over the past three decades, the known role of bile acids has expanded beyond lipid absorption to recognize their functions as signaling molecules that interact with both cell membrane (e.g., TGR5) and nuclear (e.g., FXR) receptors [1–3]. The present research focus was instigated by our observations that bile acids interact functionally with muscarinic receptors expressed by guinea pig gastric chief cells [4] and human colon cancer cells [5]. Our later finding that bile acids can activate cardiovascular muscarinic receptors [6] was confirmed by other investigators who reported that taurocholate acts as a partial agonist for neonatal rat cardiomyocyte M_2 muscarinic receptors, thereby reducing levels of intracellular cAMP and slowing the heart rate [7]. Bile acids have also been reported to interact functionally with muscarinic receptors in the lungs [8].

Cholinergic muscarinic receptors in the Class A G protein-coupled receptor family are widely expressed in mammals and play key roles in physiology and disease [9,10]. Of the five muscarinic receptor subtypes, designated M_1R-M_5R and encoded by *CHRM1–CHRM5*, the M_1R and M_3R receptor subtypes are highly expressed by intestinal epithelial cells and modulate absorption, secretion, and other important homeostatic processes. We and others demonstrated that M_3R/*CHRM3* overexpression by gastrointestinal cancers plays a functional role in tumor progression [11]. In contrast to the actions of M_3R as a colon cancer promoter, M_1R appears to function as a tumor suppressor. Strong evidence supports the latter claim—M_1R deficiency in murine colon cancer models augments colon neoplasia [12], increasing concentrations of M_1R agonists progressively reduce human colon cancer cell proliferation [13], and, compared to adjacent normal colon, M_1R/*CHRM1* expression is significantly reduced in human colon cancers [14].

Regarding the clinical implications of putative functional interactions of bile acids with cholinergic muscarinic receptors, there has long been an interest in deciphering the role of diet and gut constituents, particularly bile acids [15], as risk factors for the development and progression of gastrointestinal cancer. In this regard, the observation that feeding selected bile acids to mice can induce colon neoplasia is noteworthy [16,17]. Nonetheless, despite these interesting findings, a precise molecular basis underlying the interaction of bile acids with muscarinic receptors has been elusive. Our research has focused on filling this key knowledge gap.

Our discovery more than 27 years ago that bile acids interact functionally with muscarinic receptors [4] provided the basis for a potentially informative line of investigation. Although preliminary insights using receptor modeling to understand the structural basis for the interaction between bile acids and muscarinic receptors were instructive [18], we were puzzled by disparate results in different biological systems. Whereas atropine blocked the interaction of lithocholytaurine with guinea pig gastric chief cells [4], in other cell types, non-muscarinic receptor subtype-selective antagonists inconsistently attenuated bile acid effects. This suggested several possibilities; bile acids might not be competing with muscarinic receptor antagonists at orthotopic binding sites and/or bile acids could be interacting differently with different muscarinic receptor subtypes. Thus, to explain these discordant findings, we hypothesized that bile acids might alter muscarinic receptor activity by interacting with allosteric binding sites on specific muscarinic receptor subtypes. Whereas the potential for allosteric modulation of muscarinic receptor activity is not a novel concept [19], the recent finding of allosteric binding of cholesterol to M_1R [20] provided conceptual support for our hypothesis. As derivatives of cholesterol metabolism in the liver, bile acids share many structural features with the parent molecule [21]. Hence, it was plausible that some, if not most, functional interactions between bile acids and muscarinic receptors could be explained by their allosteric binding to M_1R.

In the present work, we used molecular modeling [22,23] to explore binding of bile acids and their glycine and taurine conjugates to M_1R. We provide in silico evidence that both primary and secondary bile acids can bind to the same allosteric site between TM5 and TM6 on M_1R as suggested for cholesterol in a previous work as well as to another allosteric site identified between TM3 and TM4. Moreover, we predict that bile acid conjugation with glycine or taurine augments allosteric binding interactions. We also analyzed the effects of bile acid binding to these predicted allosteric binding sites on M_1R activation.

2. Materials and Methods

The full computational pipeline that was conducted in the current study is illustrated in the Supporting Information (SI) Scheme S1. The workflow includes site identification by ligand competitive saturation (SILCS) simulation [24] to map fragment binding patterns on M_1R; SILCS-Hotspots analysis [25] to predict allosteric binding sites of bile acids; SILCS-Monte Carlo (SILCS-MC) docking [24] to dock acetylcholine (ACh) to the active site or bile acids to the identified allosteric binding sites; molecular dynamics (MD) simulations on the binary system of M_1R-ACh and allosterically modulated complex of M_1R-ACh-bile acids to model their dynamics; and final analyses on the MD trajectories to explore the allosteric modulations of bile acids on M_1R. Computational details are described below for each component. Chemical structures of studied molecules can be found in Supporting Information Diagram S1.

2.1. SILCS Simulations

SILCS simulations were conducted using the previously described protocol [24] to map fragment binding affinity patterns on the M_1R protein. The crystal structure of the inactive form of the M_1R protein was obtained from the protein data bank (PDB) [26] under entry 5cxv [27] with the antagonist molecule, cholesterols, lipids, and water removed, and then processed using the CHARMM-GUI server [28,29] to set up the simulation system. The protein was placed in a 1-palmitoyl-2-oleoyl-sn-glycero-3-phosphocholine (POPC) lipid bilayer to represent the membrane environment. Then 10 individual simulation systems were generated and simulated independently with each system including, in addition to the protein and bilayer, explicit water and eight randomly positioned types of solutes including benzene, propane, methanol, formamide, dimethyl ether, imidazole, methylammonium, and acetate at approximately 0.25 M each. The SILCS simulations were conducted in a set of steps composed of energy minimization, equilibration, and production runs using iterative MD/grand canonical Monte Carlo (GCMC) [30] cycles to propagate the trajectories, resulting in 20 million steps of GCMC and 100 ns MD production time per simulation, yielding a cumulative total of 200 million steps of GCMC and 1000 ns MD production time over all 10 systems. Detailed system setup and simulation methods were described in our previous work [24]. The GCMC simulations were run using the SILCS software suite, version 2023 (SilcsBio LLC, Baltimore, MD, USA), and MD simulations were performed using GROMACS (Version 4.6) [31,32], with the protein, solutes, and water being described using the CHARMM36m protein force field [33], the CHARMM General force field (CGenFF) [34], and the TIP3P water model modified for the CHARMM force field [35], respectively.

After SILCS simulations, fragment occupancy maps (FragMaps) were generated by binning selected solute or water atoms into voxels on a 1 Å spaced grid spanning the simulation system and were combined to obtain both specific and generic FragMap types as previously described [24]. The FragMaps were then normalized according to fragment bulk values under the absence of protein and were Boltzmann transformed to free energies for each functional group type to yield grid free energy (GFE) FragMaps. Four generic FragMaps were generated, including hydrophobic (APOLAR), heterocycle carbon (GEHC), hydrogen-bond donor (HBDON), and hydrogen-bond acceptor (HBACC) maps, as described previously [24]. In addition, four specific FragMaps were generated, including positive MAMN (methylammonium nitrogen) maps, negative ACEC (acetate carboxyl carbon) maps, alcohol MEOO (methanol hydroxyl oxygen) maps, and formamide carbon (FORC) maps. These FragMaps represent functional group affinity pattern-defining interactions with the protein and were used to guide the identification of allosteric binding sites and docking of bile acids to the M_1R protein. An exclusion map was also produced which describes the solute and water-forbidden regions during the SILCS simulation, thereby accounting for protein flexibility, and was used as an alternative to the protein solvent-accessible surface in the docking calculations.

2.2. SILCS-MC Docking

SILCS-MC docking was conducted to predict binding modes of ACh or bile acids to the M_1R protein. Targeting the SILCS FragMaps and the exclusion map, SILCS-MC adopts the Monte Carlo sampling method [36] to sample binding orientations and conformations of a ligand. Energy used in the Metropolis criteria relies on atomic GFE scores of selected atoms in each ligand based on a rule file that links CGenFF atom types to SILCS FragMap types, with the atomic GFE scores summed into a ligand GFE (LGFE) that is the predicted binding strength of a ligand [24]. The LGFE score along with the CGenFF intramolecular energy are combined and used as the basis of the Metropolis criteria. Non-hydrogen atoms overlapping the exclusion map are assigned a GFE value of 1000 kcal/mol that is included in the LGFE score. For the current work, SILCS-MC was conducted under an exhaustive mode where ligand conformations were sampled in the FragMaps and exclusion map by allowing the MC translational, rotational, and dihedral variations of up to 1 Å, 180 degrees, and 180 degrees at each MC step. After 10,000 steps of MC conducted at 300 K, a second round of local MC was then performed for 10,000 steps with translational, rotational, and dihedral variations of up to 0.2 Å, 9 degrees, and 60 degrees at the same temperature to further converge the sampling results. Simulated annealing (SA) [37] was then performed for 40,000 steps with the temperature gradually cooled down from 300 to 0 K with translational, rotational, and dihedral variations of up to 0.2 Å, 9 degrees, and 9 degrees to converge the sampling to a local free energy minimum. Each cycle of two MC runs and one SA run is performed up to 1000 times, with the energy difference among the top three binding poses checked every 100th cycle. If the difference is less than 0.03 kcal/mol, then the sampling is treated as being converged. The lowest energy conformation based on the LGFE is selected to represent the binding mode prediction.

2.3. SILCS-Hotspots Analyses

SILCS-Hotspots [25] calculations were conducted using SILCS FragMaps and SILCS-MC docking to globally explore potential binding sites of bile acids on the M_1R protein surface. The 3D space encompassing the full protein was divided into multiple cubic blocks, and exhaustive SILCS-MC docking was conducted for all the bile acids in each subspace with docking poses and LGFE values collected for the following clustering step. Docking poses were first clustered with respect to each bile acid to identify local binding sites. This was followed by a second round of clustering over all the bile acids to identify common binding sites. Clustering involved identifying cluster centers defined as the centroid of a ligand binding pose with the most neighbors. Following the second rounds of clustering over all the bile acids, the average LGFE score was calculated over all the bile acids in each cluster. This score was used to rank the clusters, and the highly ranked hotspot clusters were utilized to identify putative bile acid binding sites and were validated through conventional MD simulations as described in the following.

2.4. MD Simulations of ACh Bound Complexes with and Without Bile Acids

MD simulations were initiated from the crystal structure of the inactive M_1R protein (PDB ID: 5cxv) [27]. Each system included the agonist ACh at the orthosteric site obtained from SILCS-MC docking along with the best docking poses of each bile acid for the considered allosteric binding sites. The system design was selected to investigate the impact of bound bile acids on the conversion from the inactive to the active conformation. The complex structure was processed through the CHARMM-GUI server [28,29] with the membrane environment modeled using a POPC lipid bilayer, following which water and Na^+ and Cl^- ions were added, with ions at a concentration of 0.15 M. Each system was then subjected to energy minimizations followed by six rounds of MD equilibration according to the CHARMM-GUI equilibration protocol for membrane systems [38]. It should be noted that, similar to previous modeling works on muscarinic receptor proteins [39,40], we used a simplified lipid bilayer setup based on a single lipid type, POPC. Physiological membrane compositions are much more complex and joint effects from bile acid, cholesterol, and lipids could be further explored using various physiological membrane setups. Following the MD equilibration rounds, production runs were conducted in the NPT ensemble without any restraints for 1.5 μs with a 2 fs time step with periodic boundary conditions at a system temperature and pressure of 303 K and 1 atm, respectively. This temperature and pressure were controlled by the Nosé–Hoover thermostat [41,42] and Parrinello–Rahman barostat [43]. The non-bonded interactions were computed with a cutoff of 12 Å using the Verlet cutoff scheme [44]. A switching function [45] was applied for Lennard-Jones (L-J) interactions from 10 to 12 Å along with the long-range isotropic dispersion correction [46] to the energy and pressure for L-J interactions beyond 12 Å. The long-range electrostatic contribution is handled by the particle mesh Ewald (PME) method [47] with a real space cutoff of 12 Å. The same set of force fields as described above for SILCS simulation was adopted and GROMACS was used to conduct all MD simulations.

2.5. Analyses of MD Trajectories

Production run MD trajectories were analyzed using various GROMACS utilities [31,32] to extract different dynamical and structural information. Visualization of MD trajectories was performed through VMD software (Version 2.0.0) [48]. Linear mutual information (LMI) [49] analyses were conducted using the correlationplus python package [50], which was also used to make the correlation plots.

3. Results

3.1. Functional Group Binding Patterns Mapped for M_1R

SILCS was used to identify potential bile acid binding sites on M_1R. SILCS is a co-solute GCMC/MD-based methodology [24] that offers a rigorous free energy binding computational assay of utility for various molecular analyses and drug design needs [51]. In the current study, we used SILCS to predict allosteric binding sites as well as binding poses of bile acids for M_1R.

The system configuration is illustrated in Figure 1A, showing the M_1R protein embedded in a lipid bilayer with aqueous solution occupying the intra- and extracellular regions. Solute molecules are then added in the SILCS simulation to map functional group affinity patterns, termed FragMaps, of the M_1R protein. The SILCS FragMaps are shown in Figure 1B,C. As shown in Figure 1B for the entire protein, the pattern of the FragMap types corresponds to the lipid bilayer chemical environment in the central region of the protein. As shown in the green-colored meshes, apolar FragMaps are mainly found around the region of the protein interacting with the lipid bilayer while H-bond donor or acceptor or charged FragMaps are found accumulated around the intracellular and extracellular regions corresponding to the aqueous environment. Convergence of the FragMaps was verified based on the overlap of the FragMaps (overlap coefficient >0.7) from SILCS simulation runs 1–5 and 6–10 as previously discussed [24]. As additional validation for the predicted FragMap patterns for M_1R, the crystal binding mode of the antagonist Tiotropium from the inactive M_1R structure [27] used for all simulations is aligned with SILCS FragMaps at the orthosteric site (Figure 1C). The binding of the antagonist Tiotropium involves contacts between aromatic thiophene rings as well as the aliphatic nonane ring in the ligand and the hydrophobic characteristics formed by the tyrosine cage residues Y131, Y430, Y453, and Y457 as well as other surrounding aromatic or aliphatic residues such as W182, L208, and A221 [27]. Hydrogen bonding interaction between the central carbonyl group in the ligand and residue N382 and charged interaction between the quaternary ammonium ion in Tiotropium and residue D105 were also well defined. As indicated by the arrows in Figure 1C, the FragMaps at the orthosteric site reproduced important crystal binding modes of the antagonist Tiotropium, with the apolar FragMaps aligning well with the two thiophene rings, as well as the azatricyclo [3.3.1.0^{2,4}]nonane ring in the ligand and H-bond acceptor map aligning with the central carbonyl group in the ligand, and the positively charged map aligning with the quaternary ammonium ion. This validates that SILCS FragMaps accurately predicted the requirements of ligand chemical types that can be accommodated within the orthosteric site of M_1R for the particular antagonist under consideration. In addition, FragMaps also captured binding patterns that are not utilized by the current antagonist, for example, neutral H-bond donor map near the quaternary ammonium moiety which could be occupied by other antagonists, agonists, and bile acids and used to predict their binding poses.

3.2. Prediction of Allosteric Binding Sites and Binding Poses of Bile Acids

A previous work [52] studied the allosteric effect of cholesterol binding on the activation of muscarinic receptor M_1R. The work showed that binding of cholesterol slowed receptor activation. Here, we investigate the possible allosteric effect of the binding of bile acids to M_1R.

The SILCS-Hotspots workflow [25] was employed to predict potential allosteric binding sites of bile acids as well as their corresponding glycine and taurine conjugates (Table 1). Figure 2A shows the SILCS-Hotspots results with identified cluster centers (i.e., hotspots) on the protein structure and colored according to their LGFE scores. As indicated by the dashed circles, the top-ranking hotspots based on average LGFE scores are found in two regions. One is near a groove formed by intracellular halves of helices TM3 and TM4 (defined as AS34) and the second is between helices TM5 and TM6 (named AS56) on the intracellular side, respectively. The AS56 allosteric binding site resembles the previously proposed allosteric binding site for cholesterol [52].

Table 1. LGFE binding affinities (kcal/mol) for all bile acids and their glycine and taurine conjugates at the two predicted allosteric binding sites on M_1R.

Name	Acid		Glycine		Taurine	
	AS34	AS56	AS34	AS56	AS34	AS56
Cholic	−10.30	−7.41	−11.25	−8.61	−11.33	−9.75
Chenodeoxycholic	−10.64	−7.70	−11.94	−9.94	−11.53	−10.30
Deoxycholic	−9.40	−8.19	−10.99	−9.56	−11.29	−9.95
Lithocholic	−9.91	−7.48	−12.30	−10.29	−12.12	−10.72

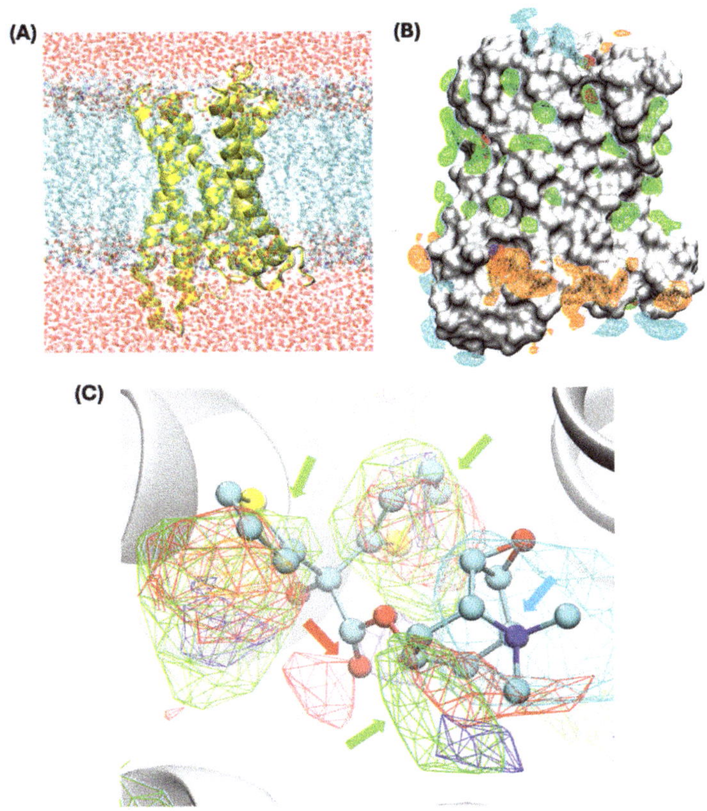

Figure 1. Simulation system and SILCS FragMaps surrounding the full protein and in the active site. (**A**) The main simulation system used for the current study. The M_1R protein (yellow) is embedded in a lipid bilayer composed of POPC molecules (shown in ball and stick representation with carbons in cyan); intra- and extracellular regions were modeled using water (only oxygen atoms are shown in red) and Na^+ and Cl^- ions (not shown). (**B**) SILCS FragMaps for the M_1R protein are shown for apolar (green), hydrogen-bond acceptor (red) and donor (blue), and negative (orange) and positive (cyan) functional group classes rendered at GFE levels of −1.0, −1.0, −1.0, −1.2, and −1.2 kcal/mol, respectively. The protein is shown in the surface representation. (**C**) Zoomed-in view of FragMaps at the orthosteric site aligned with the crystal binding orientation of the antagonist Tiotropium. Protein is shown in the cartoon representation. FragMaps are rendered with the same levels and colors as in panel (**B**). Consistency of the FragMap types and the chemical groups in the antagonist are indicated by arrows using colors consistent with the FragMap type color code.

The predicted binding mode of lithocholic acid as well as that of its glycine conjugate for sites AS34 and AS56 are shown in Figure 2B. The binding modes of the other molecules can be found in Figures S1 and S2 in the Supporting Information. Binding of lithocholic acid at the AS34 site is predicted to be stabilized by the steroid skeleton's contacts with the hydrophobic environment created by residues L118, F121, F125, Y133, and M145, as also indicated by the apolar FragMaps shown in green meshes. Salt bridge interactions between the tail carboxylate group in lithocholic acid and basic residues R129 and K136 are also present, as indicated by the negatively charged FragMaps shown as orange meshes. For site AS56, hydrophobic contacts between the steroid skeleton in bile acids and neighboring residues L207, I211, Y212, A364, and L367, as well as salt bridge interactions between the carboxylate tail and K361 and R365, are seen. Similar binding modes for other bile acids at the two predicted allosteric binding sites are observed, indicating the consistency of bile acid binding to M_1R. With increasing tail length, the predicted binding poses of glycine and taurine conjugates can go deeper into the lipid bilayer region and have contacts with more hydrophobic residues at the top of the binding groove. For example, lithocholylglycine can have hydrophobic contacts with more residues such as M114, L117, L148, A149, and V152 at the AS34 site, and M204, C205, Y208, L371, L372, and I375 for the AS56 site. And the carboxylate tail can form a salt bridge interaction with alternative basic residues on the intracellular side, such as R141 for the AS34 site and R123 for the AS56 site. Again, similar binding modes for other bile acid conjugates are seen as presented in the Supporting Information.

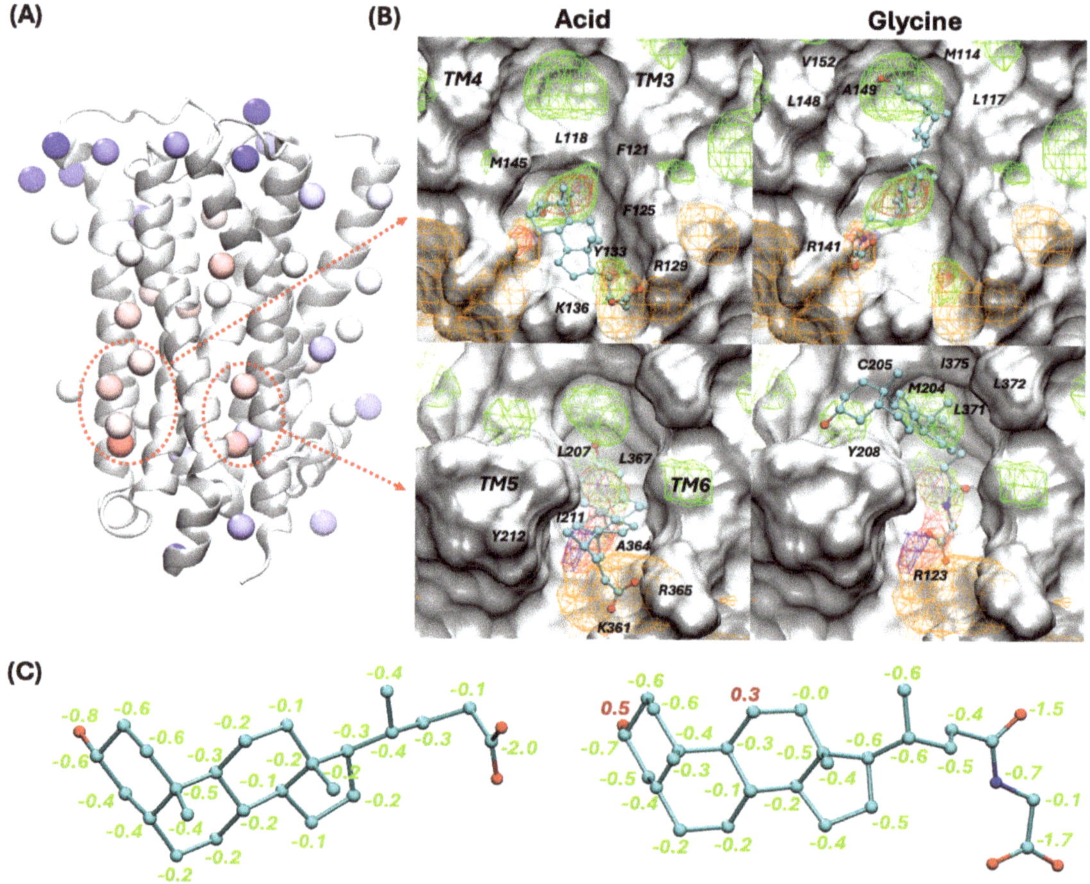

Figure 2. SILCS-Hotspots analysis and predicted docking poses of a representative bile acid and its glycine conjugate at the two top-ranked binding sites. (**A**) SILCS-Hotspots showing the predicted bile acid hotspot sites. Binding cluster centers are shown in spheres and colored in the red–white–blue scale for LGFE scores, with red indicating more favorable LGFE values. The two top-ranked binding sites are indicated by red dashed circles (AS34 on the left and AS56 on the right). (**B**) Docking poses of lithocholic acid and its glycine conjugate at the two predicted binding grooves. The protein surface and SILCS FragMaps at the same rendering level as Figure 1 are shown. Important binding site residues are labeled. (**C**) Atom GFE contributions to the total LGFE score for lithocholic acid and its glycine conjugate at the AS34 binding groove.

Table 1 lists the predicted binding affinities in terms of LGFEs for all ligands considering both allosteric sites. Predicted binding affinities are similar across different types of bile acids for the individual sites. Upon conjugation with glycine and taurine, binding is generally enhanced. Comparing the two identified allosteric binding sites, all bile acids as well as their conjugates favor site AS34 over AS56. Figure 2C shows the atomic GFE contributions of each classified atom to the overall predicted binding affinity in terms of LGFE for lithocholic acid. The aliphatic atoms throughout the steroid skeleton each make small favorable contributions consistent with their overlap with the apolar FragMaps, making the largest overall contribution to binding. Favorable contributions are also observed from the polar atoms in the sidechains, with the individual acid groups making substantially favorable contributions. GFE contribution analyses for other molecules can be found in Table S1; similar distributions for the different functional groups are seen. Regarding glycine and taurine bile acid conjugates, the introduction of glycine and taurine groups allows the molecule to achieve an ideal size to occupy other hydrophobic subsites in the binding groove while maintaining other favorable polar interactions and leading to significantly enhanced overall binding affinities. For example, for lithocholylglycine, shown in the right panel in Figure 2C, the glycine portion contributes −4.0 kcal/mol GFE in total vs. −2.0 kcal/mol for the carboxylate group alone without conjugation. To compare our findings to the previous work on cholesterol, we also docked cholesterol to the two binding sites, yielding LGFE values for the AS34 and AS56 sites of −10.68 and −7.10 kcal/mol, respectively, which are comparable to our predicted bile acid binding affinities. Moreover, bile acid glycine or taurine conjugates are predicted to bind more tightly to M_1R at both allosteric sites than cholesterol. In addition to the docking results, in the future, MD simulations on systems containing both bile acid and cholesterol could be explored to study their competition and their joint effects on modulating M_1R.

3.3. Lithocholic Bile Acids Stay in the Two Predicted Binding Sites Through Microsecond MD on the M1R-ACh-Bile Acid Complexes

MD simulations were undertaken to verify the stability of the lithocholic bile acids in the binding sites identified by SILCS-MC and their potential allosteric impact on receptor activation. The simulations were initiated from the inactive M_1R crystal structure [27] with the physiological muscarinic receptor agonist ACh docked to the orthosteric binding site using SILCS-MC. The simulation setup was selected based on the assumption that the bile acids will bind to and stabilize the inactive form of M_1R as they are negative allosteric modulators even in the presence of an agonist. This setup also allows for the impact of the bile acids on conformational transitions between the inactive and active forms to be monitored, as presented in Section 3.4 below. The three lithocholic molecules, including both their bile acid and conjugate forms, were selected based on their highly favorable SILCS-MC LGFE scores (Table 1), with the docked orientations used to initiate the simulations. This includes simulations of the M_1R-ACh-bile acid allosterically modulated complexes with the bile acids docked in either of the two top-ranked allosteric binding sites, AS34 and AS56.

Figure 3 plots the distance between the COM of lithocholylglycine during the MD simulation and that of the initial docking pose for both allosteric sites. The relatively stable COM distance values across the simulation, with fluctuations for both sites, indicate the ligand is accommodated well in the binding sites, though with frequent reorientations. The same plots for lithocholic acid and its taurine conjugate can be found in Figure S3. Figure 3 also shows the binding poses of lithocholylglycine from the final frame of the MD simulation for the two sites, and the binding poses are comparable to the docking poses shown in Figure 2B, even though reorientation of the steroid skeleton and carboxylate tail is seen.

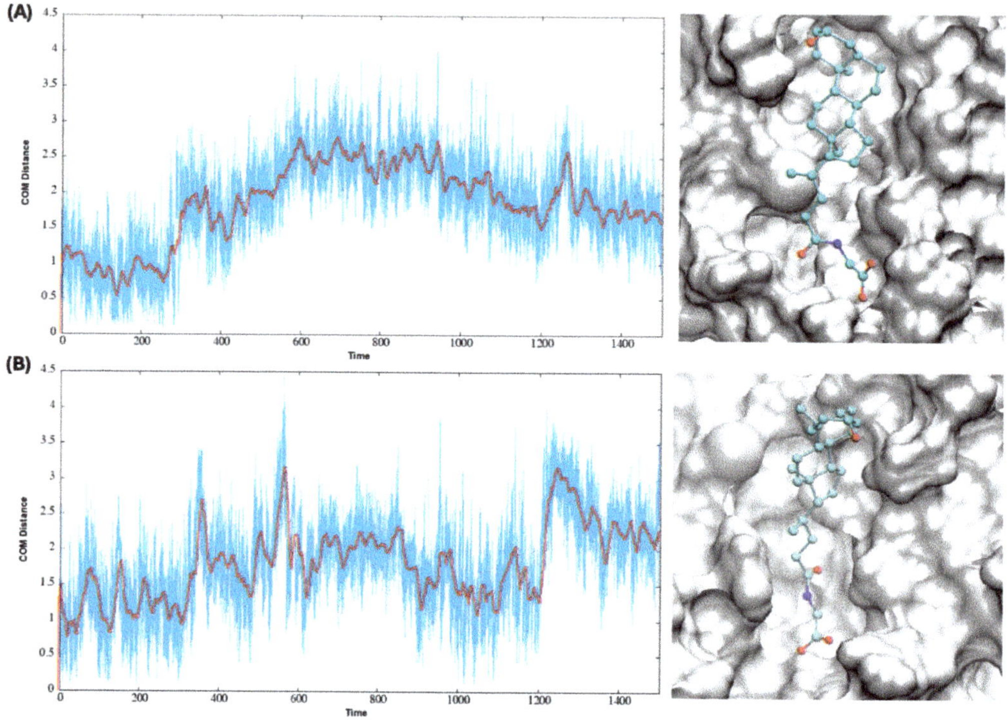

Figure 3. Center-of-mass (COM) distance (in Å) between lithocholylglycine during the MD simulations (time in ns) and its initial docking orientation and the binding poses from the last frame of the MD simulations for sites AS34 (**A**) and AS56 (**B**). Moving average values of distances calculated every 15 ns are also shown as red lines.

Figure 4 shows the interaction analysis of lithocholylglycine with binding site residues from the MD simulation. At the AS34 site, lithocholylglycine is in close contact with the selected hydrophobic residues on TM3 and TM4 shown in Figure 2B for most of the simulation time, with close contacts with residues L118, F125, and M145 for almost 100% of the simulation time. Close contacts of lithocholylglycine with basic residues are also seen associated with the carboxylate tail, potentially forming a salt bridge interaction with those residues, especially R129. Similarly, for the AS56 site, the ligand stays close to most of the labeled hydrophobic residues on TM5 and TM6, with close contacts formed with residues Y208 and L371 for almost all the simulation time. Again, close contacts between the ligand with basic residues in the binding site are seen, with R123 being the most frequent residue in contact. The results are similar for lithocholic acid and taurine conjugate and can be found in Figure S4. The relatively stable binding of bile acids observed through the microsecond MD simulations further validates the two identified allosteric binding sites for bile acids. The predicted interacting residues with bile acids at the allosteric sites could help to design mutagenesis studies to experimentally validate such bile acid binding predictions. For example, residues F125 and R129 were predicted to serve as key nonpolar and polar interacting residues, respectively, with lithocholylglycine at the AS34 site. Accordingly, site-directed mutations such as F125A and R129Q could be employed to experimentally validate bile acid binding and the associated allosteric effects predicted for this site.

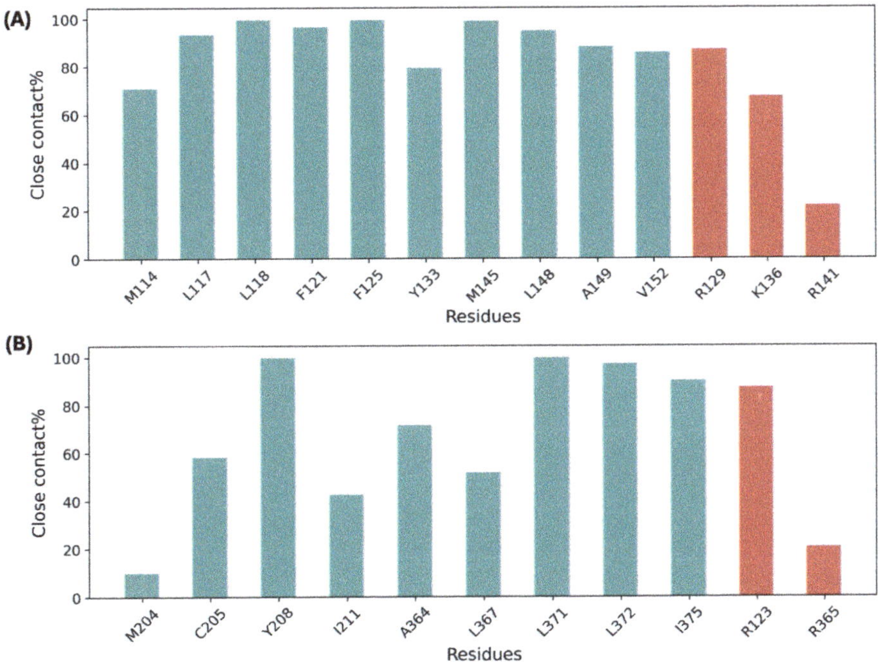

Figure 4. Percentages of close contact formed by lithocholylglycine with binding site residues along the whole MD simulation for sites AS34 (**A**) and AS56 (**B**). Close contact is defined as any non-hydrogen atom in the ligand being within 3 Å of any non-hydrogen atom in the protein residue. Hydrophobic residues are colored in green and basic residues in red.

3.4. Transition Toward the Active State upon Agonist Binding with and Without the Presence of Bile Acids

Given the experimental evidence that bile acids and cholesterol act as negative allosteric modulators of M_1R, the allosterically modulated complex MD simulations were analyzed to observe the impact of the presence of selected bile acids on the transition from the inactive to the active conformation. As a control, a simulation of the M_1R-ACh binary complex was performed to determine the extent of conformational change from the inactive form toward the active form in the absence of the selected bile acids.

Previous structural studies on activation of muscarinic receptors show multiple important conformational changes between inactive and active states. These include the global conformational changes involving the outward movement of TM6 and inward movement of TM5 on the cytoplasmic end for the active state and the residue interaction network on the cytoplasmic end undergoing large changes upon activation [39,40,53,54]. The ion lock defined as the salt bridge between two charged residues on TM3 and TM6 (TM3:R123-TM6:E360 for M_1R), which is present in the inactive state, is broken in the active state. Formation of a hydrogen bond between two tyrosine residues on TM5 and TM7 (TM5:Y208-TM7:Y418 for M_1R) involving the NPxxY motif is also observed for the active state. Another conformational indicator for the active state is the DRY motif in TM3, as previously reported [39,40]. The DRY motif includes R123, which is involved in the ionic lock mentioned above.

Analysis of the MD simulations focused on the conformational changes occurring during the MD simulations as well as on the binding orientation of ACh, which was shown to remain in the orthosteric site (Figure S5). Figure 5A shows the root-mean-square deviation (RMSD) of the protein backbone atoms during the M_1R-ACh binary complex MD simulation, and a climbing trend is observed which indicates conformational changes are happening associated with the activation process. In Figure 5B, the final structure from the 1.5 μs MD simulation is aligned with both the inactive (PDB entry:5cxv) [27] and the active (PDB entry:6oij) [55] crystal structures of M_1R. The outward tilting of TM6 and inward tilting of TM5 on the cytoplasmic end compared to the inactive structure are evident, though the changes have not reached the full active conformation. The two tyrosine residues in the final frame of the MD simulation are much closer to each other when compared to the original inactive structure but are still blocked by L367 and have not yet formed a hydrogen bond.

To quantify the inactive to active conformational change during the MD simulation, we tracked the RMSD of the NPxxY motif and the distance between residues R123 and E360 involved in the ionic lock. A two-dimensional analysis of the RMSD of the NPxxY motif versus the COM distance between the two ion lock residues R123 and E360 as a function of simulation time is shown in Figure 5C. The figure includes the values observed in the active and inactive crystal structures. As can be seen, in the M_1R-ACh binary complex simulation, the two descriptors shift from their positions in the inactive state (starting structure as indicated by the red point on the X-axis) toward those of the active state (indicated by the red star symbol) by the end of the simulation. This includes the distortion of the NPxxY motif as well as the separation of the ion lock residues, indicating the binary complex MD simulation captures, to a large extent, conformational changes associated with activation of M_1R upon the binding of agonist ACh. The distortion of the NPxxY motif happens very quickly and it reached a stable conformation around 0.8 μs, following which the separation of the two ionic lock residues began to occur. To fully observe the transition to the active conformation, a longer MD simulation or MD simulations with enhanced sampling—for example, the accelerated MD method as used previously [39,40]—would be required, but since our focus was to study the potential allosteric effects of bile acids, the extent of the conformational change that occurred during the 1.5 μs conventional MD simulation was deemed appropriate.

Figure 5D,E show the two geometric metrics, the COM distance between R123 and E360 versus the RMSD of the NPxxY motif as a function of the simulation time for the allosterically modulated complex M_1R AS34- and AS56-bound lithocholylglycine systems. As compared to the bile acid-absent scenario shown in Figure 5C, despite distortions of the NPxxY motif, short distances between the two ionic residues were maintained for both the AS34- and AS56-bound lithocholylglycine systems in the μs-long MD simulations. This indicates that the conformational transition from inactive to active state caused by binding of ACh to the orthosteric site was hindered upon binding of lithocholylglycine to both suggested allosteric binding sites on M_1R. As shown in Figure S6, similar trends were observed for lithocholic acid and lithocholyltaurine.

3.5. Activation Mechanism of M_1R upon ACh Binding

Previous studies of other muscarinic receptor proteins proposed multiple activation mechanisms with different levels of atomic detail [39,54]. Miao et al. applied the dual-boost accelerated MD method to study M_2R activation [39]. Starting from an inactive conformation in the absence of the antagonist, they sampled broad collections of conformations of the apo protein and suggested an allosteric activation pathway that contains two typical intermediate states. During the transition from the inactive state to intermediate state 1, active site residues W400 on TM6 and Y430 on TM7, as well as residue Y440 on the TM7 intracellular side, underwent conformational changes. On the path from the intermediate state 1 to 2, Y206 on the TM5 intracellular side reorients its sidechain from the initial position between TM3 and TM6 to the lipid-exposed side of TM6. During the final transition to the active state, Y206 and Y440 relocate their sidechains toward each other to form a hydrogen bond. The study found that W400 is a key residue acting as a conformational switch, consistent with the transmission switch role of this conserved tryptophan residue found for adenosine receptors [39]. The study also emphasized the role of active site residue Y430, which is involved in a key conformational change involving its relocation from the active site.

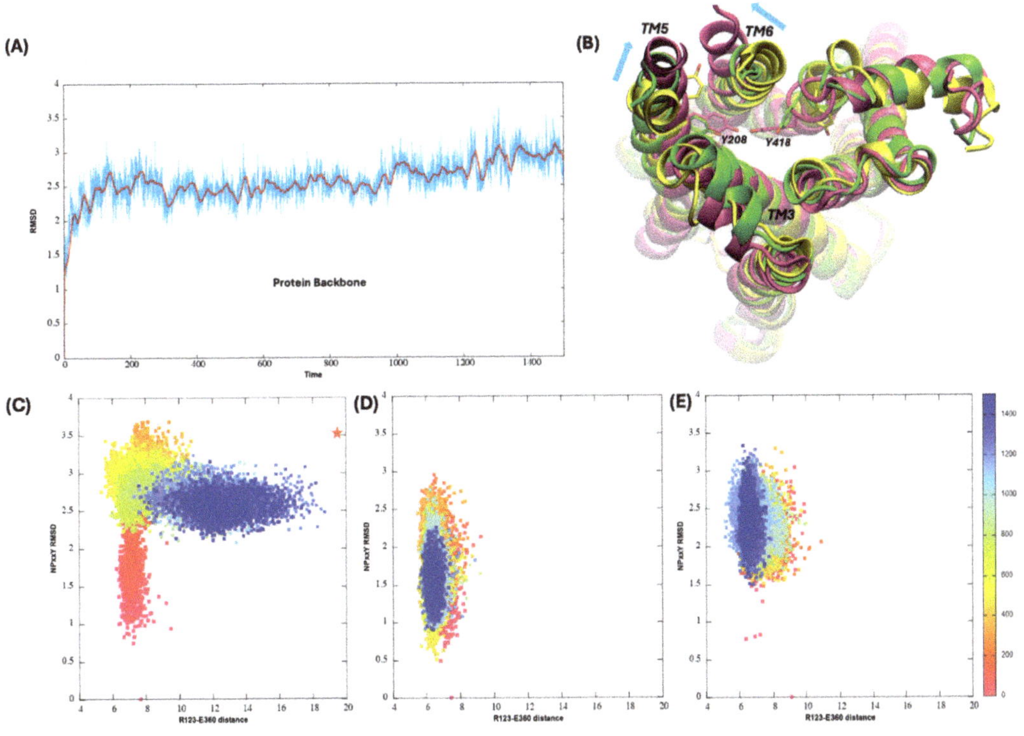

Figure 5. Analyses of the MD trajectories for the M_1R-ACh binary complex and the allosterically modulated complex M_1R-ACh-lithocholylglycine. (**A**) RMSD in Å of protein backbone atoms during the M_1R-ACh binary complex production MD simulation, with time shown in ns. Moving average values of RMSD calculated every 15 ns are also shown as red lines. (**B**) Comparison of active (PDB: 6oij in purple) and inactive (PDB:5cxv in yellow) crystal structures as well as the last frame protein structure (green) from MD on the binary complex. Residues Y208 and Y418 are shown. (**C**) Two-dimensional plot of COM distance between residues R123 and E360 versus the RMSD of the NPxxY motif with reference to the inactive conformation along the MD simulation on the binary complex. The red star symbol indicates the values from the crystal structure of the active M_1R. The same two-dimensional plots from MD simulations on the allosterically modulated complex M_1R-ACh-lithocholylglycine are also shown for both the AS34 site (**D**) and the AS56 site (**E**) binding of the bile acid glycine conjugate.

In another study on M_2R, Gong et al. used MD simulations to study the activation process due to the agonists ACh and Iperoxo [54]. In that work, they focused on the conformational properties of residues on TM5 and TM6 and suggested a detailed conformational coupling pathway in transducing signals across the membrane from the active site to the intracellular side. Again, the transmission switch role of W400 was confirmed and its coupling with F396 and following sequential conformational dynamics through residues L393 and I389 along TM6 were suggested.

The trajectory from the present MD simulation on the binary M_1R-ACh complex was examined frame-by-frame, and key conformational changes related to the activation were identified. The orthosteric site is composed of the so-called tyrosine cage formed by Y106 on TM3, Y381 on TM6, and Y404 and Y408 on TM7, along with W378 on TM6 and D105 on TM3. Upon ACh binding, the interaction network will affect the orthosteric site residue dynamics, leading to conformational changes down the TM helices. Three important sequential conformational changes that can be linked to activation were observed in the MD trajectory. First, there was distortion of the NPxxY motif, especially conformational changes in Y418 and its hydrogen bonding network. Second, disruption of the ionic lock caused by the structural dynamics of W378 as well as Y418 involved hydrogen bonding interactions with R123. Third, the structural dynamics of residue Y208 changed. These observations for M_1R are consistent with previous findings for M_2R but differ in atomic details. Detailed analysis can be found in the SI, and it should be noted that this observation is based on a single trajectory conducted in our study as a control for other allosterically modulated complex MD simulations; multiple MD simulations could be conducted on the binary system to further validate such observations.

3.6. Bile Acid Binding Stabilized the Allosteric Sites and Other TM Regions

To further explore how bile acid allosteric binding dampens the activation of M_1R, we analyzed residue flexibility in terms of root-mean-square fluctuation (RMSF). Figure 6A shows the RMSF values of all residues from MD simulations on M_1R-ACh, compared with simulations performed on M_1R-ACh-lithocholylglycine complexes considering both the AS34 and AS56 sites. In general, when compared to simulation without the bile acid, less residue flexibility throughout the full protein is observed for both AS34- and AS56-bound lithocholylglycine. As expected, the AS34-bound system shows the lowest RMSF values for the TM3 and TM4 regions among the three simulations while the AS56-bound system shows smaller RMSF values for most parts of TM5 and TM6 than the other two simulations. Similar trends are observed for lithocholic acid and its taurine conjugate systems, as shown in Figure S7. Figure 6B shows the RMSF distributions focusing on allosteric site residues for both the AS34 and AS56 sites. Compared to the bile acid-absent case, the more concentrated distributions toward smaller RMSF values for both AS34- and AS56-bound systems at both sites AS34 and AS56 indicate that bile acid binding stabilized the allosteric sites.

3.7. Allosteric Site Stabilization Weakened Cooperativity Between TM Domains

We also analyzed the dynamic couplings between residues in M_1R in the MD simulation by calculating a linear correlation metric in terms of LMI of protein structure fluctuations [50]. LMI was suggested to be a better metric than other linear correlation metrics such as dynamic cross-correlation, which was adopted in a previous work on muscarinic receptors [39], since it avoids the angular dependency issue [50]. Figure 7A shows the dynamic map of residue LMI correlations, and a medium level of correlations is observed in many places on the off-diagonal regions which indicates strong dynamic couplings between residues from different TM domains across the entire protein during the transition process. Focusing on regions with LMI values larger than 0.625, five off-diagonal regions as indicated by red dashed circles, designated 1–5, are observed and listed in Table 2. These noted regions are related to the correlated motions between residues on TM1 and TM2, TM3 and TM4, TM5 and TM6, TM6 and TM7, and TM1 and TM7, respectively. Such highly correlated motions between TM5 and TM6 and TM6 and TM7 were also observed previously in Miao et al.'s study on the M_2R activation process [39]. This indicates that the binding of ACh triggered high correlations of motions of different TM domains, and such a strong correlation network facilitated the conformational transition from the inactive to active state.

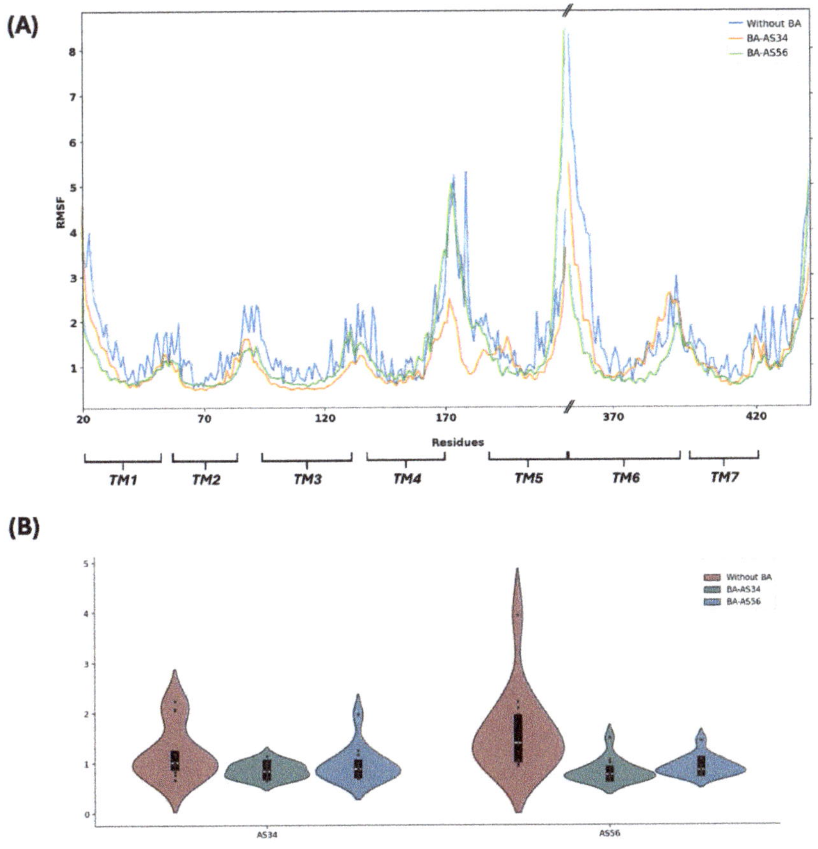

Figure 6. Root-mean-square fluctuation analyses of MD simulations with and without bile acid. (**A**) RMSFs of all residues in M_1R through MD simulations of M_1R-ACh and M_1R-ACh-lithocholylglycine considering both the AS34 and AS56 sites. (**B**) Distribution of RMSF values of key residues at sites AS34 and AS56 as defined in Figure 2B. The mean value is indicated as the white horizontal line and the black bar defines the third and first quartiles.

Figure 7B,C present the LMI correlation plots for the AS34- and AS56-bound lithocholylglycine complexes. For AS34 site-bound simulation, different correlation patterns are seen when compared to the bile acid absence case. For the AS56 site-bound scenario, fewer correlations are observed compared to Figure 7A, suggesting a disruption of the network between TM domains, which may be critical for activation. Focusing on the five highly correlated regions identified in the bile acid-absent system, for both the AS34 and AS56 cases, weakened correlations are observed for off-diagonal areas 3 and 4 that describe correlated motions for TM5-TM6 and TM6-TM7. This indicates that for both the AS34 and AS56 allosteric site-bound lithocholylglycine systems, the dynamic cooperativity between TM5 and TM6 and between TM6 and TM7 observed in the bile acid-absent system is diminished. Similar trends are observed for lithocholic acid and its taurine conjugates, as shown in Figure S8. As discussed above, the proposed activation mechanism suggests that the interaction network between residues on TM5, TM6, and TM7 is essential for facilitating collective conformational changes that lead to activation. Thus, the enhanced allosteric site stability upon bile acid binding reduces the flexibility of various TM domain residues and weakens cooperativity between TM domains, particularly between the TM5–TM6 and TM6–TM7 pairs, ultimately slowing activation.

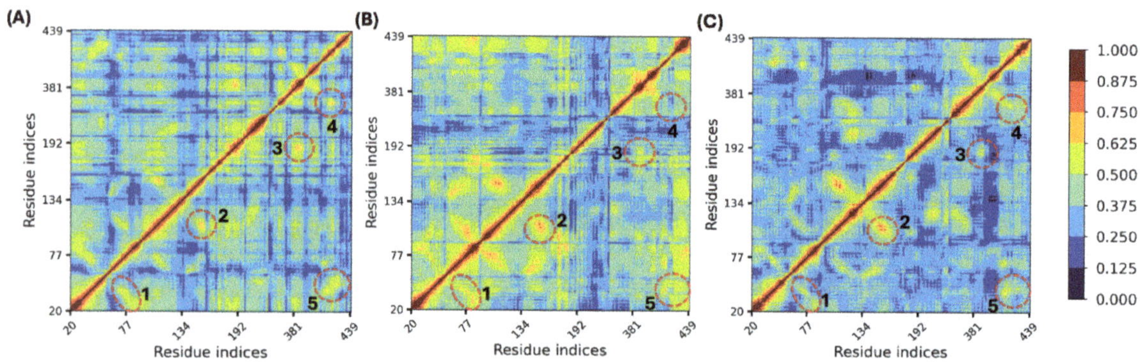

Figure 7. Linear mutual information residue correlation plots calculated from MD simulations. (**A**) LMI for the M_1R-ACh complex and the M_1R-ACh-lithocholylglycine complex targeting both the (**B**) AS34 and (**C**) AS56 allosteric sites. Five off-diagonal regions with high LMI values identified for M_1R-ACh complex simulation are indicated by red dashed circles. Detailed interaction information is tabulated in Table 2.

Table 2. Residue pairs that have high LMI values identified in the five off-diagonal regions as shown in Figure 7.

Off-Diagonal Region	Domain 1		Domain 2	
	TM	Residues	TM	Residues
1	TM1	27/29/31	TM2	83
		37/40		75/76
		39/40/41/43		71/72
2	TM3	105–114	TM4	151–157
		99		161
3	TM5	183–191	TM6	383–390
4	TM6	364–370	TM7	416–421
5	TM1	39	TM7	412
		41–43		414–416
		48/49		425/427/428
		55/56		424/425

3.8. Allosteric Bile Acid Binding Distorts Agonist ACh Binding

Binding of bile acid at allosteric sites not only degraded cooperativity of TM domains but also distorted binding of agonist ACh at the active site. Figure 8A shows the RMSD plots for non-hydrogen atoms of ACh along the MD simulation time for the AS34 and

AS56 allosteric site-bound lithocholylglycine systems. When compared with the RMSD plot for the bile acid absence case as shown in Figure S5, much larger fluctuations are seen for ACh when bile acids are present considering both allosteric sites. Figure 8B shows the close contact percentages of ACh with each key residue that defines the active site through the simulation. When bile acid is absent, ACh has close contacts with multiple residues at the active site for almost the full MD simulation time, especially with D105, Y106, W378, Y404, and Y408. ACh also forms a charged interaction between the quaternary basic nitrogen in ACh and D105 that is crucial for its binding. However, a much-weakened charged interaction between ACh and D105 is observed for both AS34- and AS56-bound lithocholylglycine systems, as indicated by the smaller contact percentage values. Less stable contacts of ACh with most tyrosine residues as well as W378 are also observed for both AS34- and AS56-bound lithocholylglycine cases, though to different extents. Interestingly, enhanced contacts of ACh with Y381 are observed when lithocholylglycine is present. Results for lithocholic acid and its taurine conjugate can be found in Figure S9, and the overall trend of weakened contacts of ACh with active site residues, especially with D105 and W378, is confirmed upon the allosteric binding of various bile acid forms. As discussed above, W378 is a key transmission switch responsible for conformational dynamics on TM6 needed for transition to active conformation; loose contact of ACh with W378 is expected to disrupt the activation process.

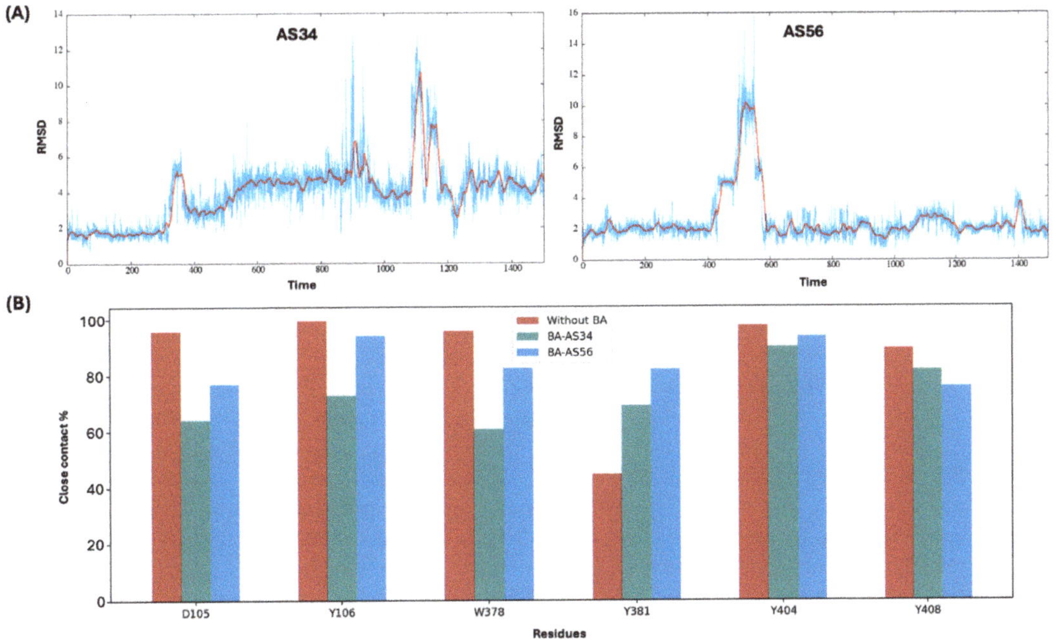

Figure 8. Analyses of ACh binding at the active site. (**A**) RMSD (in Å) of ACh heavy atoms along the MD simulation (in ns) for AS34- and AS56-bound lithocholylglycine systems. Moving average values of distances calculated every 15 ns are also shown as red lines. (**B**) Percentage of close contact formed by ACh with active site residues along the whole MD simulation for M_1R-ACh and M_1R-ACh-lithocholylglycine systems for both the AS34 and AS56 sites.

4. Discussion

In the current work, we employed SILCS and MD simulations to explore bile acid binding to M_1R. We used SILCS functional group binding patterns on M_1R to predict allosteric binding sites of bile acids as well as their binding poses. As reported previously for cholesterol, bile acids were predicted to bind to the same allosteric site on M_1R between TM5 and TM6. We also identified another potential allosteric site, AS34 between TM3 and TM4, which can accommodate the bile acid better than the AS56 site. Showing that lithocholic acid and its glycine and taurine conjugates can stay in the two suggested allosteric sites during the full microsecond MD simulations provides additional proof of concept.

Using SILCS-Monte Carlo simulations, the endogenous agonist ACh was docked to the active site of a M_1R inactive conformation, and the M_1R activation process was captured using a 1.5 μs MD simulation. The activation mechanism of M_1R upon agonist binding determined using the MD trajectory is consistent with the proposed mechanisms for other muscarinic receptors. In parallel, MD simulations on bile acid-bound M_1R-ACh complexes were conducted with the extracted dynamic conformational information compared to the bile acid-absent scenario. It was observed that bile acid binding slows the activation process of M_1R and cooperativity between TM domains; binding between TM5 and TM6 and between TM6 and TM7 was especially weakened. These findings suggest that bile acids can allosterically modulate M_1R activation by enhancing the structural stability of TM3/TM4 or TM5/TM6, leading to a loss of the cooperativity between TM5, TM6, and TM7 needed for activation. Moreover, distortions of ACh binding in the active site observed in MD trajectories when bile acid is present further support the allosteric modulation of M_1R activity by bile acids.

It should be emphasized that the current work is purely based on computational modeling predictions and needs to be thoroughly validated using wet lab methods, for example, mutagenesis, to validate the allosteric binding sites we propose. It has not escaped our attention that given the broad distribution of muscarinic receptors, including all the segments of the gastrointestinal tract, and that bile acids circulate in the bloodstream, the potential role for bile acids regulating cell function via allosteric effects on M_1R and other muscarinic receptor subtype signaling in health and disease is exhaustive. Previous studies revealed that bile acid concentrations in the proximal colon are in the high micromolar range, levels that could result in negative allosteric modulation of M_1R activity and, thereby, alter the function of both normal and neoplastic colonocytes via this mechanism [56].

Considering the structural similarities among muscarinic receptors, it is anticipated that cholesterol and bile acid could share similar allosteric modulating mechanisms with muscarinic receptor subtypes beyond M_1R. Moreover, after experimental validation, the current in silico findings suggest the possibility that bile acids interact functionally with other G protein-coupled receptors, acting as allosteric modulators. In this context, others highlighted the therapeutic potential of leveraging allosteric interactions to modulate muscarinic receptor activity and function [57]. We plan to explore these prospects in future work. The potential impact of our findings is augmented by the current focus on developing new drugs that selectively stimulate or block M_1R activation [58] and advances in understanding the role that gut bacteria play in modulating the diversity of human bile acids [59].

5. Conclusions

Using the novel SILCS method as well as conventional MD simulations, conformational dynamics of agonist bound M_1R with and without the presence of bile acids or their glycine and taurine conjugates were fully modeled and analyzed. The simulation observations indicate the potential for unconjugated and conjugated bile acids to serve as negative allosteric modulators of M_1 muscarinic receptors. In the context of colon cancer, where M_1R activation is reported to suppress neoplastic cell proliferation, this suggests a novel mechanism whereby bile acids can promote colon neoplasia. More broadly, our finding that allosteric binding of bile acids to G protein-coupled receptors may modulate receptor function is worthy of further exploration.

Supplementary Materials: The following supporting information can be downloaded at https://www.mdpi.com/article/10.3390/biom15091326/s1: Scheme S1: Flowchart illustrating the computational pipeline conducted in the current work; Diagram S1: Two-dimensional chemical structures of all small molecules studied in the current work; Figure S1: Predicted binding poses for all considered bile acid and glycine and taurine conjugates at the AS34 site using SILCS-MC; Figure S2: Predicted binding poses for all considered bile acid and glycine and taurine conjugates at the AS56 site using SILCS-MC; Table S1: Functional group GFE contributions to LGFE (kcal/mol) for all bile acids and their glycine and taurine conjugates at the two predicted allosteric binding sites on M_1R; Figure S3: COM distance between the COM of lithocholic acid or lithocholyltaurine during the MD simulations and that of the initial docking orientation and the binding pose from the last frame of the MD simulations for sites AS34 and AS56; Figure S4: Percentage of close contact formed by lithocholic acid or lithocholyltaurine with binding site residues along the whole MD simulation for sites AS34 and AS56; Figure S5: RMSD of ACh heavy atoms during the M_1R-ACh binary complex MD simulation and COM distance between the COM of ACh during the MD simulations and that of the initial docking orientation; Figure S6: Two-dimensional plot of COM distance between residues R123 and E360 versus the RMSD of the NPxxY motif with reference to the inactive conformation along the MD simulation for site AS34 and AS56 binding of lithocholic acid and lithocholyltaurine; Figure S7: RMSFs of all residues in M_1R through MD simulations of M_1R-ACh and M_1R-ACh-lithocholic acid or M_1R-ACh-lithocholyltaurine considering both the AS34 and AS56 sites; Figure S8: LMI residue correlation plots calculated from MD simulations on M_1R-ACh-lithocholic acid and the M_1R-ACh-lithocholyltaurine complex targeting both the AS34 and AS56 allosteric sites; Figure S9: RMSD of ACh heavy atoms along the MD simulation for AS34- and AS56-bound lithocholic acid or lithocholyltaurine systems; atomic level of detail observed in the MD trajectory for activation of M_1R.

Author Contributions: Conceptualization, W.Y., J.-P.R., A.D.M.J. and D.J.W.; formal analysis, W.Y. and A.D.M.J.; investigation, W.Y. and A.D.M.J.; data curation, W.Y. and A.D.M.J.; writing—original draft preparation, W.Y. and J.-P.R.; writing—review and editing, W.Y., J.-P.R., A.D.M.J. and D.J.W.; funding acquisition, J.-P.R., A.D.M.J. and D.J.W. All authors have read and agreed to the published version of the manuscript.

Funding: This work was supported by Merit Review Award BX004890 from the United States (U.S.) Department of Veterans Affairs Biomedical Laboratory Research and Development Program (to J-P. R.). The contents do not represent the views of the U.S. Department of Veterans Affairs or the United States Government. Financial support from NIH grants R35 GM131710 (to A.D.M.) and R01 GM129327 (to D.J.W.) and computational support from the University of Maryland Computer-Aided Drug Design Center are also acknowledged.

Institutional Review Board Statement: Not applicable.

Informed Consent Statement: Not applicable.

Data Availability Statement: The original contributions presented in this study are included in the article/supplementary material. Further inquiries can be directed to the corresponding author(s).

Conflicts of Interest: ADM is the co-founder and CSO of SilcsBio LLC.

Abbreviations

The following abbreviations are used in this manuscript:

M_1R	M_1 muscarinic receptor
ACh	Acetylcholine
POPC	1-palmitoyl-2-oleoyl-sn-glycero-3-phosphocholine
SILCS	Site identification by ligand competitive saturation
MD	Molecular dynamics
FEP	Free energy perturbation
LGFE	Ligand grid free energy
MC	Monte Carlo
PDB	Protein data bank
LMI	Linear mutual information
RMSD	Root-mean-square deviation
RMSF	Root-mean-square fluctuation
COM	Center-of-mass

References

1. Perino, A.; Demagny, H.; Velazquez-Villegas, L.; Schoonjans, K. Molecular Physiology of Bile Acid Signaling in Health, Disease, and Aging. *Physiol. Rev.* **2021**, *101*, 683–731. [CrossRef]
2. Fleishman, J.S.; Kumar, S. Bile Acid Metabolism and Signaling in Health and Disease: Molecular Mechanisms and Therapeutic Targets. *Signal Transduct. Target. Ther.* **2024**, *9*, 97. [CrossRef] [PubMed]
3. Kiriyama, Y.; Nochi, H. The Biosynthesis, Signaling, and Neurological Functions of Bile Acids. *Biomolecules* **2019**, *9*, 232. [CrossRef] [PubMed]
4. Raufman, J.-P.; Zimniak, P.; Bartoszko-Malik, A. Lithocholyltaurine Interacts with Cholinergic Receptors on Dispersed Chief Cells from Guinea Pig Stomach. *Am. J. Physiol. Gastrointest. Liver Physiol.* **1998**, *274*, G997–G1004. [CrossRef]
5. Cheng, K.; Chen, Y.; Zimniak, P.; Raufman, J.-P.; Xiao, Y.; Frucht, H. Functional Interaction of Lithocholic Acid Conjugates with M3 Muscarinic Receptors on a Human Colon Cancer Cell Line. *Biochim. Biophys. Acta (BBA) Mol. Basis Dis.* **2002**, *1588*, 48–55. [CrossRef]
6. Khurana, S.; Raufman, J.-P.; Pallone, T.L. Bile Acids Regulate Cardiovascular Function. *Clin. Transl. Sci.* **2011**, *4*, 210–218. [CrossRef]
7. Sheikh Abdul Kadir, S.H.; Miragoli, M.; Abu-Hayyeh, S.; Moshkov, A.V.; Xie, Q.; Keitel, V.; Nikolaev, V.O.; Williamson, C.; Gorelik, J. Bile Acid-Induced Arrhythmia Is Mediated by Muscarinic M2 Receptors in Neonatal Rat Cardiomyocytes. *PLoS ONE* **2010**, *5*, e9689. [CrossRef]
8. Urso, A.; D'Ovidio, F.; Xu, D.; Emala, C.W.; Bunnett, N.W.; Perez-Zoghbi, J.F. Bile Acids Inhibit Cholinergic Constriction in Proximal and Peripheral Airways from Humans and Rodents. *Am. J. Physiol. Lung Cell. Mol. Physiol.* **2020**, *318*, L264–L275. [CrossRef]
9. Kruse, A.C.; Hu, J.; Kobilka, B.K.; Wess, J. Muscarinic Acetylcholine Receptor X-Ray Structures: Potential Implications for Drug Development. *Curr. Opin. Pharmacol.* **2014**, *16*, 24–30. [CrossRef]
10. Kruse, A.C.; Kobilka, B.K.; Gautam, D.; Sexton, P.M.; Christopoulos, A.; Wess, J. Muscarinic Acetylcholine Receptors: Novel Opportunities for Drug Development. *Nat. Rev. Drug Discov.* **2014**, *13*, 549–560. [CrossRef]
11. Cheng, K.; Shang, A.C.; Drachenberg, C.B.; Zhan, M.; Raufman, J.-P. Differential Expression of M3 Muscarinic Receptors in Progressive Colon Neoplasia and Metastasis. *Oncotarget* **2017**, *8*, 21106–21114. [CrossRef]
12. Cheng, K.; Xie, G.; Khurana, S.; Heath, J.; Drachenberg, C.B.; Timmons, J.; Shah, N.; Raufman, J.-P. Divergent Effects of Muscarinic Receptor Subtype Gene Ablation on Murine Colon Tumorigenesis Reveals Association of M3R and Zinc Finger Protein 277 Expression in Colon Neoplasia. *Mol. Cancer* **2014**, *13*, 77. [CrossRef] [PubMed]
13. Sundel, M.H.; Sampaio Moura, N.; Cheng, K.; Chatain, O.; Hu, S.; Drachenberg, C.B.; Xie, G.; Raufman, J.-P. Selective Activation of M1 Muscarinic Receptors Attenuates Human Colon Cancer Cell Proliferation. *Cancers* **2023**, *15*, 4766. [CrossRef] [PubMed]
14. Alizadeh, M.; Schledwitz, A.; Cheng, K.; Raufman, J.-P. Mechanistic Clues Provided by Concurrent Changes in the Expression of Genes Encoding the M1 Muscarinic Receptor, β-Catenin Signaling Proteins, and Downstream Targets in Adenocarcinomas of the Colon. *Front. Physiol.* **2022**, *13*, 857563. [CrossRef] [PubMed]
15. Glinghammar, B.; Rafter, J. Carcinogenesis in the Colon: Interaction between Luminal Factors and Genetic Factors. *Eur. J. Cancer Prev.* **1999**, *8*, S87–S94.

16. Bernstein, C.; Holubec, H.; Bhattacharyya, A.K.; Nguyen, H.; Payne, C.M.; Zaitlin, B.; Bernstein, H. Carcinogenicity of Deoxycholate, a Secondary Bile Acid. *Arch. Toxicol.* **2011**, *85*, 863–871. [CrossRef]
17. Bernstein, H.; Bernstein, C. Bile Acids as Carcinogens in the Colon and at Other Sites in the Gastrointestinal System. *Exp. Biol. Med.* **2023**, *248*, 79–89. [CrossRef]
18. Raufman, J.-P.; Chen, Y.; Cheng, K.; Compadre, C.; Compadre, L.; Zimniak, P. Selective Interaction of Bile Acids with Muscarinic Receptors: A Case of Molecular Mimicry. *Eur. J. Pharmacol.* **2002**, *457*, 77–84. [CrossRef]
19. Jakubík, J.; El-Fakahany, E.E. Allosteric Modulation of Muscarinic Acetylcholine Receptors. *Pharmaceuticals* **2010**, *3*, 2838–2860. [CrossRef]
20. Jakubík, J.; El-Fakahany, E.E. Allosteric Modulation of GPCRs of Class A by Cholesterol. *Int. J. Mol. Sci.* **2021**, *22*, 1953. [CrossRef]
21. Russell, D.W. Fifty Years of Advances in Bile Acid Synthesis and Metabolism. *J. Lipid Res.* **2009**, *50*, S120–S125. [CrossRef]
22. Schlick, T. *Molecular Modeling and Simulation: An Interdisciplinary Guide*; Springer: New York, NY, USA, 2010; Volume 2.
23. Hollingsworth, S.A.; Dror, R.O. Molecular Dynamics Simulation for All. *Neuron* **2018**, *99*, 1129–1143. [CrossRef] [PubMed]
24. Goel, H.; Hazel, A.; Ustach, V.D.; Jo, S.; Yu, W.; MacKerell, A.D. Rapid and Accurate Estimation of Protein–Ligand Relative Binding Affinities Using Site-Identification by Ligand Competitive Saturation. *Chem. Sci.* **2021**, *12*, 8844–8858. [CrossRef] [PubMed]
25. MacKerell, A.D.; Jo, S.; Lakkaraju, S.K.; Lind, C.; Yu, W. Identification and Characterization of Fragment Binding Sites for Allosteric Ligand Design Using the Site Identification by Ligand Competitive Saturation Hotspots Approach (SILCS-Hotspots). *Biochim. Biophys. Acta (BBA) Gen. Subj.* **2020**, *1864*, 129519. [CrossRef] [PubMed]
26. Sussman, J.L.; Lin, D.; Jiang, J.; Manning, N.O.; Prilusky, J.; Ritter, O.; Abola, E.E. Protein Data Bank (PDB): Database of Three-Dimensional Structural Information of Biological Macromolecules. *Acta Crystallogr. Sect. D* **1998**, *54*, 1078–1084. [CrossRef]
27. Thal, D.M.; Sun, B.; Feng, D.; Nawaratne, V.; Leach, K.; Felder, C.C.; Bures, M.G.; Evans, D.A.; Weis, W.I.; Bachhawat, P.; et al. Crystal Structures of the M1 and M4 Muscarinic Acetylcholine Receptors. *Nature* **2016**, *531*, 335–340. [CrossRef]
28. Jo, S.; Kim, T.; Iyer, V.G.; Im, W. CHARMM-GUI: A Web-Based Graphical User Interface for CHARMM. *J. Comput. Chem.* **2008**, *29*, 1859–1865. [CrossRef]
29. Jo, S.; Cheng, X.; Lee, J.; Kim, S.; Park, S.-J.; Patel, D.S.; Beaven, A.H.; Lee, K.I.; Rui, H.; Park, S.; et al. CHARMM-GUI 10 Years for Biomolecular Modeling and Simulation. *J. Comput. Chem.* **2017**, *38*, 1114–1124. [CrossRef]
30. Mezei, M. A Cavity-Biased (T, V, μ) Monte Carlo Method for the Computer Simulation of Fluids. *Mol. Phys.* **1980**, *40*, 901–906. [CrossRef]
31. Van Der Spoel, D.; Lindahl, E.; Hess, B.; Groenhof, G.; Mark, A.E.; Berendsen, H.J.C. GROMACS: Fast, Flexible, and Free. *J. Comput. Chem.* **2005**, *26*, 1701–1718. [CrossRef]
32. Abraham, M.J.; Murtola, T.; Schulz, R.; Páll, S.; Smith, J.C.; Hess, B.; Lindahl, E. GROMACS: High Performance Molecular Simulations through Multi-Level Parallelism from Laptops to Supercomputers. *SoftwareX* **2015**, *1–2*, 19–25. [CrossRef]
33. Huang, J.; Rauscher, S.; Nawrocki, G.; Ran, T.; Feig, M.; de Groot, B.L.; Grubmüller, H.; MacKerell, A.D. CHARMM36m: An Improved Force Field for Folded and Intrinsically Disordered Proteins. *Nat. Methods* **2017**, *14*, 71–73. [CrossRef] [PubMed]
34. Yu, W.; He, X.; Vanommeslaeghe, K.; MacKerell, A.D., Jr. Extension of the CHARMM General Force Field to Sulfonyl-Containing Compounds and Its Utility in Biomolecular Simulations. *J. Comput. Chem.* **2012**, *33*, 2451–2468. [CrossRef] [PubMed]
35. Durell, S.R.; Brooks, B.R.; Ben-Naim, A. Solvent-Induced Forces between Two Hydrophilic Groups. *J. Phys. Chem.* **1994**, *98*, 2198–2202. [CrossRef]
36. Metropolis, N.; Ulam, S. The Monte Carlo Method. *J. Am. Stat. Assoc.* **1949**, *44*, 335–341. [CrossRef]
37. Kirkpatrick, S.; Gelatt, C.D.; Vecchi, M.P. Optimization by Simulated Annealing. *Science* **1983**, *220*, 671–680. [CrossRef]
38. Wu, E.L.; Cheng, X.; Jo, S.; Rui, H.; Song, K.C.; Dávila-Contreras, E.M.; Qi, Y.; Lee, J.; Monje-Galvan, V.; Venable, R.M.; et al. CHARMM-GUI Membrane Builder toward Realistic Biological Membrane Simulations. *J. Comput. Chem.* **2014**, *35*, 1997–2004. [CrossRef]
39. Miao, Y.; Nichols, S.E.; Gasper, P.M.; Metzger, V.T.; McCammon, J.A. Activation and Dynamic Network of the M2 Muscarinic Receptor. *Proc. Natl. Acad. Sci. USA* **2013**, *110*, 10982–10987. [CrossRef]
40. Miao, Y.; Caliman, A.D.; McCammon, J.A. Allosteric Effects of Sodium Ion Binding on Activation of the M3 Muscarinic G-Protein-Coupled Receptor. *Biophys. J.* **2015**, *108*, 1796–1806. [CrossRef]
41. Nosé, S. A Unified Formulation of the Constant Temperature Molecular Dynamics Methods. *J. Chem. Phys.* **1984**, *81*, 511–519. [CrossRef]
42. Hoover, W.G. Canonical Dynamics: Equilibrium Phase-Space Distributions. *Phys. Rev. A* **1985**, *31*, 1695–1697. [CrossRef]
43. Parrinello, M.; Rahman, A. Polymorphic Transitions in Single Crystals: A New Molecular Dynamics Method. *J. Appl. Phys.* **1981**, *52*, 7182–7190. [CrossRef]

44. Páll, S.; Hess, B. A Flexible Algorithm for Calculating Pair Interactions on SIMD Architectures. *Comput. Phys. Commun.* **2013**, *184*, 2641–2650. [CrossRef]
45. Steinbach, P.J.; Brooks, B.R. New Spherical-Cutoff Methods for Long-Range Forces in Macromolecular Simulation. *J. Comput. Chem.* **1994**, *15*, 667–683. [CrossRef]
46. Allen, M.P.; Tildesley, D.J. *Computer Simulation of Liquids*; Oxford University Press: Oxford, UK, 2017; ISBN 0-19-252470-4.
47. Essmann, U.; Perera, L.; Berkowitz, M.L.; Darden, T.; Lee, H.; Pedersen, L.G. A Smooth Particle Mesh Ewald Method. *J. Chem. Phys.* **1995**, *103*, 8577–8593. [CrossRef]
48. Humphrey, W.; Dalke, A.; Schulten, K. VMD: Visual Molecular Dynamics. *J. Mol. Graph.* **1996**, *14*, 33–38. [CrossRef]
49. Lange, O.F.; Grubmüller, H. Generalized Correlation for Biomolecular Dynamics. *Proteins Struct. Funct. Bioinform.* **2006**, *62*, 1053–1061. [CrossRef]
50. Tekpinar, M.; Neron, B.; Delarue, M. Extracting Dynamical Correlations and Identifying Key Residues for Allosteric Communication in Proteins by Correlationplus. *J. Chem. Inf. Model.* **2021**, *61*, 4832–4838. [CrossRef]
51. Yu, W.; Weber, D.J.; MacKerell, A.D. Computer-Aided Drug Design: An Update. In *Antibiotics: Methods and Protocols*; Sass, P., Ed.; Springer: New York, NY, USA, 2023; pp. 123–152; ISBN 978-1-0716-2855-3.
52. Randáková, A.; Dolejší, E.; Rudajev, V.; Zimčík, P.; Doležal, V.; El-Fakahany, E.E.; Jakubík, J. Role of Membrane Cholesterol in Differential Sensitivity of Muscarinic Receptor Subtypes to Persistently Bound Xanomeline. *Neuropharmacology* **2018**, *133*, 129–144. [CrossRef]
53. Miao, Y.; McCammon, J.A. Graded Activation and Free Energy Landscapes of a Muscarinic G-Protein–Coupled Receptor. *Proc. Natl. Acad. Sci. USA* **2016**, *113*, 12162–12167. [CrossRef]
54. Gong, Z.; Zhang, X.; Liu, M.; Jin, C.; Hu, Y. Visualizing the M2 Muscarinic Acetylcholine Receptor Activation Regulated by Aromatic Ring Dynamics. *bioRxiv* **2024**, bioRxiv:2024.02.18.580854. [CrossRef]
55. Maeda, S.; Qu, Q.; Robertson, M.J.; Skiniotis, G.; Kobilka, B.K. Structures of the M1 and M2 Muscarinic Acetylcholine Receptor/G-Protein Complexes. *Science* **2019**, *364*, 552–557. [CrossRef] [PubMed]
56. Hamilton, J.P.; Xie, G.; Raufman, J.-P.; Hogan, S.; Griffin, T.L.; Packard, C.A.; Chatfield, D.A.; Hagey, L.R.; Steinbach, J.H.; Hofmann, A.F. Human Cecal Bile Acids: Concentration and Spectrum. *Am. J. Physiol. Gastrointest. Liver Physiol.* **2007**, *293*, G256–G263. [CrossRef]
57. Zhu, L.; Rossi, M.; Cohen, A.; Pham, J.; Zheng, H.; Dattaroy, D.; Mukaibo, T.; Melvin, J.E.; Langel, J.L.; Hattar, S.; et al. Allosteric Modulation of β-Cell M3 Muscarinic Acetylcholine Receptors Greatly Improves Glucose Homeostasis in Lean and Obese Mice. *Proc. Natl. Acad. Sci. USA* **2019**, *116*, 18684–18690. [CrossRef]
58. Tobin, A.B. A Golden Age of Muscarinic Acetylcholine Receptor Modulation in Neurological Diseases. *Nat. Rev. Drug Discov.* **2024**, *23*, 743–758. [CrossRef]
59. Guzior, D.V.; Okros, M.; Shivel, M.; Armwald, B.; Bridges, C.; Fu, Y.; Martin, C.; Schilmiller, A.L.; Miller, W.M.; Ziegler, K.M.; et al. Bile Salt Hydrolase Acyltransferase Activity Expands Bile Acid Diversity. *Nature* **2024**, *626*, 852–858. [CrossRef]

biomolecules MDPI

Article

Conditioned Generative Modeling of Molecular Glues: A Realistic AI Approach for Synthesizable Drug-like Molecules

Naeyma N. Islam [1] and Thomas R. Caulfield [2,*]

1 Department of Neuroscience, Mayo Clinic, Jacksonville, FL 32224, USA; islam.naeyma@mayo.edu
2 Digital Ether Computing, Miami, FL33130, USA
* Correspondence: thomas@digitalethercomputing.com

Abstract: Alzheimer's disease (AD) is marked by the pathological accumulation of amyloid beta-42 (Aβ42), contributing to synaptic dysfunction and neurodegeneration. While extracellular amyloid plaques are well-studied, increasing evidence highlights intracellular Aβ42 as an early and toxic driver of disease progression. In this study, we present a novel, Generative AI–based drug design approach to promote targeted degradation of Aβ42 via the ubiquitin–proteasome system (UPS), using E3 ligase–directed molecular glues. We systematically evaluated the ternary complex formation potential of Aβ42 with three E3 ligases (CRBN, VHL, and MDM2) through structure-based modeling, ADMET screening, and docking. We then developed a Ligase-Conditioned Junction Tree Variational Autoencoder (LC-JT-VAE) to generate ligase-specific small molecules, incorporating protein sequence embeddings and torsional angle-aware molecular graphs. Our results demonstrate that this generative model can produce chemically valid, novel, and target-specific molecular glues capable of facilitating Aβ42 degradation. This integrated approach offers a promising framework for designing UPS-targeted therapies for neurodegenerative diseases.

Keywords: molecular glues; artificial intelligence; variable autoencoders; ligase-conditioned junction tree variational autoencoder

1. Introduction

Alzheimer's disease (AD), the most prevalent neurodegenerative disorder, is characterized by progressive cognitive decline and neuronal loss. The accumulation of amyloid beta 42 (Aβ42) peptides, derived from the aberrant cleavage of amyloid precursor protein (APP) by β- and γ-secretases, is a central pathological hallmark [1–4]. Aβ42 is particularly aggregation-prone and toxic compared to Aβ40, forming intracellular deposits that precede extracellular amyloid plaques and correlate with early synaptic dysfunction and neuronal death [1,5–8]. Thus, targeting intracellular Aβ42 is crucial for developing effective AD therapies.

Despite numerous therapeutic efforts—including secretase inhibition, immunotherapy, and enhancing proteostasis—significant challenges persist. Secretase inhibitors suffer from off-target toxicities due to the broad substrate specificity of γ- and β-secretases [9–11]. Immunotherapeutic strategies, while promising, face issues with blood–brain barrier (BBB) penetration and potential CNS inflammation [12,13]. Enhancement of autophagy to clear Aβ42 aggregates has shown potential but remains nonspecific and may compromise essential cellular functions [14].

The ubiquitin–proteasome system (UPS) offers an attractive alternative for degrading intracellular Aβ42 aggregates [15,16]. However, Aβ42's aggregation-prone nature impedes its efficient recognition by the UPS. Molecular glues—small molecules that induce or stabilize interactions between E3 ligases and target proteins—have emerged as a promising strategy to facilitate Aβ42 degradation through ubiquitination and proteasomal clearance [15–19]. Unlike PROTACs, molecular glues require no bifunctional linker and can enhance BBB permeability due to their smaller size [17–19].

Selecting the appropriate E3 ligase is critical. VHL, CRBN, and MDM2 are prominent candidates based on their structural properties and prior success in targeted protein degradation [20–24]. VHL, associated with hypoxia signaling; CRBN, known for ligand-dependent substrate recruitment; and MDM2, a versatile p53 regulator, offer complementary substrate specificities that could be leveraged for Aβ42 targeting. In this work, we evaluate these three E3 ligases for their potential to form stable ternary complexes with Aβ42 and design molecular glues.

Traditional molecular glue discovery relies on extensive chemical library screening, making it slow and resource-intensive. Artificial intelligence (AI) models now provide a powerful alternative, enabling efficient de novo design, lead optimization, and virtual screening [25–30]. Several generative AI models have been employed for small molecule design; however, each presents notable limitations. Sequence-based models, such as those relying on SMILES representations, often suffer from non-uniqueness and syntactic errors that lead to chemically invalid molecules, complicating the learning process and limiting their reliability in drug discovery [31]. Similarly, standard Variational Autoencoders (VAEs) and Generative Adversarial Networks (GANs), although effective at sampling chemical space, often fail to guarantee chemical validity and structural coherence in the generated compounds [32,33].

To overcome these challenges, we selected the Junction Tree Variational Autoencoder (JTVAE) [34–37], a graph-based model that encodes molecular structures as hierarchical trees of chemically meaningful substructures, enabling the generation of valid, drug-like molecules with superior structural fidelity. Unlike sequence-based or flat graph models, JTVAE preserves chemical rules during molecule generation, making it particularly well-suited for drug design applications. While the original JTVAE framework models molecular structures as 2D graphs, it does not account for torsional flexibility, a critical factor for molecular stability, bioactivity, and ligand–receptor interactions. To address this limitation, we extended JTVAE by explicitly integrating torsional angles into bond representations, enabling the model to capture conformational dynamics and generate structurally diverse, chemically realistic molecules with improved 3D predictive fidelity. Moreover, to design molecular glues specific to different E3 ligases, we further enhanced the model by incorporating protein-level information. Specifically, we embedded the amino acid sequences of binding site residues from CRBN, VHL, and MDM2 into the training pipeline. This allowed the model to learn ligase-specific binding patterns, enabling conditional generation of molecular glues optimized for forming stable ternary complexes between Aβ42 and individual E3 ligases. Together, these advancements substantially improve the chemical validity, conformational realism, and biological specificity of the molecules generated compared to the original JTVAE framework.

In this study, we investigate the potential of three widely studied E3 ubiquitin ligases—VHL, CRBN, and MDM2—for mediating the targeted degradation of amyloid beta (Aβ42) through molecular glue strategies. Furthermore, we design ligase-specific novel molecular glues by incorporating protein sequence embeddings derived from the binding site residues of each E3 ligase into a Junction Tree Variational Autoencoder (JT-VAE), a graph-based generative model. To enhance the precision of molecular glue generation, we also integrate three-dimensional conformational features—including molecular rotational angles—alongside conventional structural and chemical descriptors. This multi-modal approach allows for the generation of chemically valid, drug-like small molecules that are optimized for selective engagement with E3 ligases and effective degradation of Aβ42.

2. Methods

2.1. Amyloid Beta 42 Structure

For structural studies of Aβ42, which corresponds to residues 627–713 of the full-length APP, we selected the cryo-EM structure of Type I Aβ42 filaments from human brain tissue (PDB ID: 7Q4B) [38]. These fibrils, resolved at 2.5 Å, represent a homo-decameric assembly of Aβ42 peptides adopting an S-shaped conformation, a structural feature crucial to their aggregation and stability in Alzheimer's disease pathology [39]. Although 7Q4B depicts extracellular fibrils, they originate from intracellular oligomerization processes, making this structure highly relevant to the pathological cascade [40].

The raw PDB structure was preprocessed using Schrödinger's Protein Preparation Wizard in Maestro [41]. Preprocessing steps included bond order assignment, addition of missing hydrogen atoms, removal of water molecules beyond 5 Å, protonation state adjustment with Epik at pH 7.0 ± 2.0, and optimization of the hydrogen bonding network. A restrained energy minimization was performed with the OPLS4 force field, allowing hydrogen atoms to relax while constraining heavy atoms. The resulting prepared structure was subsequently used for all in silico analyses.

2.2. Binding Site Identified in Amyloid Beta42

To identify ligandable regions on the Aβ42 fibril structure, we employed the SiteMap module within Schrödinger Maestro to predict and characterize potential binding pockets based on physicochemical and geometric criteria [41].

2.3. E3 Ligase Structure Selection and Preparation

High-quality crystallographic structures of three E3 ligases—VHL (PDB ID: 4W9C), CRBN (PDB ID: 4CI3), and MDM2 (PDB ID: 4OAS)—were selected based on resolution and relevance to small-molecule recruitment. The VHL structure (1.45 Å) offers a well-defined ligand-binding interface suitable for docking studies targeting Aβ42 degradation. The CRBN structure, resolved with thalidomide, reveals the canonical molecular glue binding site, while the MDM2 structure features a co-crystallized inhibitor occupying the druggable p53-interacting cleft. All structures were prepared using Schrödinger's Protein Preparation Wizard [41], with preprocessing steps including removal of non-essential heteroatoms and waters, bond order assignment, hydrogen addition, hydrogen bonding optimization, and restrained energy minimization using the OPLS3e or OPLS4 force fields. These curated models served as the structural foundation for E3 ligase–specific molecular glue design. Prepared structures of VHL, CRBN, and MDM2 are shown in Figure 1.

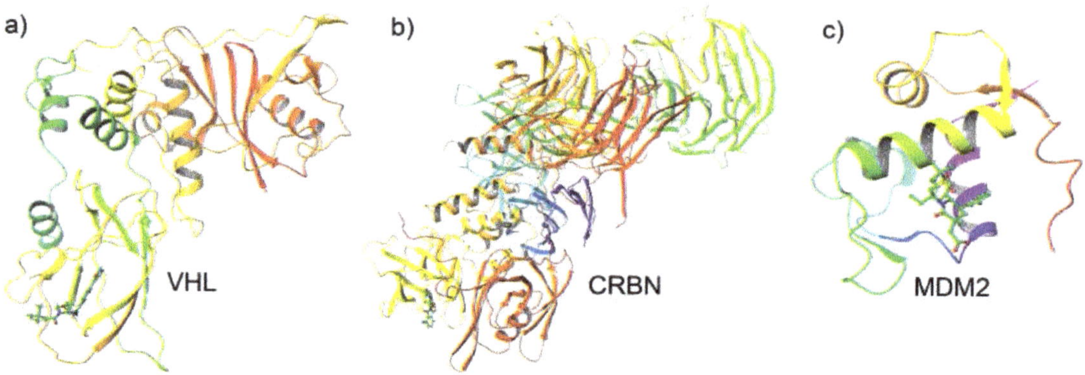

Figure 1. Prepared 3D structures of representative E3 ligases used in this study, highlighting bound ligands in green stick representation. (**a**) Von Hippel–Lindau (VHL), (**b**) Cereblon (CRBN), and (**c**) Mouse Double Minute 2 homolog (MDM2). Each structure is visualized with ligands occupying their respective binding pockets, confirming appropriate docking orientation and active site integrity.

2.4. ADMET-Based Filtering and Docking

To optimize drug-likeness and safety, in silico ADMET (Absorption, Distribution, Metabolism, Excretion, and Toxicity) screening was performed on compound libraries from the Vitas and ChEMBL databases [42,43] using Schrödinger's QikProp and ADMET/Tox modules [44]. Compounds were filtered based on Lipinski's Rule of Five and Veber's rule, followed by predictions for intestinal absorption, blood–brain barrier permeability, CYP450 inhibition, hepatotoxicity, and AMES mutagenicity. Only candidates meeting all pharmacokinetic and toxicity thresholds were retained for downstream studies. Filtered compounds were docked into the ligand-binding sites of VHL, CRBN, and MDM2 using Schrödinger's Glide module in standard precision (SP) and extra precision (XP) modes. For each docked complex, Aβ42 was aligned near the E3-ligase interface, and ternary complexes were refined using Rosetta FlexPepDock and evaluated based on docking scores.

Additionally, all-atom molecular dynamics (MDs) simulations were conducted using the Desmond simulation engine (Schrödinger Suite) with the OPLS4 force field. Ternary complexes were solvated in TIP3P water, neutralized with counterions, minimized, equilibrated under NPT conditions (300 K, 1 atm) using the Berendsen thermostat and barostat, and subjected to 100 ns production runs.

2.5. Dataset Construction and Feature Engineering for Ligase-Conditioned Molecular Glue Design

The training dataset consisted of ADMET-screened and docking-prioritized compounds capable of forming ternary complexes with CRBN, VHL, and MDM2 E3 ligases and amyloid-β42. All molecules were converted into molecular graph and junction tree formats using RDKit and the original JT-VAE preprocessing pipeline located at https://github.com/croningp/JTNN-VAE (accessed on 15 November 2024), which includes subdirectories for fast_molvae and python script preprocess.py at https://github.com/croningp/JTNN-VAE/blob/master/fast_molvae/preprocess.py (accessed on 15 November 2020). To enhance molecular representation beyond the capabilities of the original JT-VAE framework, we extended the preprocessing pipeline to incorporate torsional (dihedral) angle features for rotatable bonds. These dihedral angles were computed using a custom RDKit-based routine, which identifies rotatable bonds and calculates their dihedral values via the custom Python function "get_torsional_angle" (Supplementary Figure S1). This addition enables the model to better capture conformational flexibility in small molecules. Furthermore, each compound was paired with the corresponding E3 ligase's binding site sequence embedding, allowing the model to learn protein-contextual chemical features. Together, these enhancements—dihedral angle integration and protein sequence conditioning—enabled the development of a modified JT-VAE framework capable of generating E3 ligase-specific molecular glues for CRBN, VHL, and MDM2.

2.6. Development of Ligase-Conditioned JT-VAE Incorporating Torsional Features for Molecular Glue Design

To enable the generation of E3 ligase–specific molecular glues, we extended the canonical Junction Tree Variational Autoencoder (JT-VAE) framework introduced by Jin et al. [45], originally designed for chemically valid molecule generation without target protein context [46]. While the standard JT-VAE effectively encodes molecular graphs and junction trees, it lacks two critical capabilities for structure-guided drug design: (i) conditioning molecule generation on protein-specific information, and (ii) modeling three-dimensional (3D) conformational flexibility via torsional (dihedral) angles.

To address these limitations, we developed a Ligase-Conditioned JT-VAE (LC-JT-VAE). First, we enhanced the molecular graph representation by incorporating torsional angle features into the bond-level message passing steps. Dihedral angles were computed using RDKit's GetDihedralDeg function by identifying rotatable bonds (Supplementary Figure S1), providing critical 3D information on conformational dynamics essential for accurate ligand design. Second, we integrated sequence embeddings derived from the binding site residues of CRBN, VHL, and MDM2 directly into the molecular latent space, enabling the model to generate ligase-specific molecular glues.

2.7. Encoding E3 Ligase Binding Site Sequences

Binding site residue sequences for CRBN, VHL, and MDM2 were encoded through multiple strategies to capture structural and functional contexts. Initially, one-hot encoding and k-mer encoding were applied to extract local sequence patterns. For more advanced, context-aware representations, ProtBERT—a transformer model pretrained on UniRef100—was utilized to generate high-dimensional embeddings [47]. These embeddings were further refined using bidirectional long short-term memory (BiLSTM) networks [48] to model sequential dependencies, and then projected to fixed-length vectors (e.g., 128D or 512D) for compatibility with the molecular latent space, thus enabling protein-conditioned molecule generation.

2.8. LC-JT-VAE Architecture and Latent Space Fusion

The LC-JT-VAE architecture consists of two parallel encoders: a molecular graph encoder, comprising a message-passing neural network (MPN) and a junction tree message-passing network (JTMPN), and a protein sequence encoder for E3 ligase binding site embeddings. These two streams were fused in the latent space using either concatenation with linear projection and ReLU activation, or cross-attention mechanisms where the molecular graph attended to the protein embedding. The fused latent vector was passed to a conditional Junction Tree Neural Network (JTNN) decoder, which reconstructed the junction tree scaffold and molecular graph conditioned on the ligase binding context (Figure 2).

Implementation summary:

```
# During encoding
mol_latent = JTNNEncoder(mol_graph, junction_tree)
ligase_embed = LigaseEncoder(sequence)
```

```
# Latent fusion
z = torch.cat([mol_latent, ligase_embed], dim=1)
# During decoding
output = JTNNDecoder(z)
```

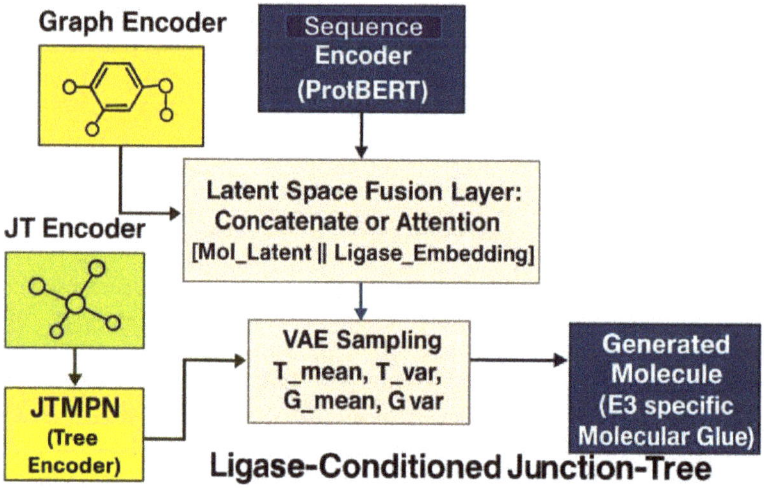

Figure 2. Conditional JTNNVAE architecture for generating E3 ligase-specific molecular glues. Molecular graphs are encoded via MPN, and junction trees via JTMPN, then passed to the JTNNEncoder. Simultaneously, binding site sequences from E3 ligases (e.g., CRBN, VHL, MDM2) are encoded into vector representations. These embeddings are fused to form a conditioned latent space, which is sampled using VAE parameters (T_mean, T_var, G_mean, G_var). The conditional JTNNDecoder reconstructs a chemically valid molecule tailored to the specific ligase context.

2.9. Detailed Architectural Breakdown and Conditioning Mechanism

- **Molecular Graph and Junction Tree Encoding**: Molecules were processed as graphs and junction trees using MPN and JTMPN. Bond features were enhanced by incorporating torsional angles computed via RDKit, improving the model's ability to capture conformational flexibility crucial for protein–ligand interactions.
- **Ligase Sequence Encoding**: Binding site sequences were embedded using ProtBERT-based encoders and projected to match the molecular latent space dimensions.
- **Latent Space Fusion Strategies**:
    - Concatenation and Linear Projection:
      z_fused = ReLU(W[z_mol; z_seq] + b)
    - Cross-Attention Mechanism:
      z_mol′ = MultiHeadAttention(z_mol, z_seq, z_seq)
- Conditional Decoding:
    - Tree Decoder: Reconstructs the junction tree scaffold.
    - Graph Decoder: Rebuilds the full molecular graph conditioned on the fused latent vector.

The detailed internal architecture of LC-JT-VAE, including the JTNNEncoder, Graph-GRU layers, sequence encoder modules (W_H1, W_H2, W_O), and latent projections (T_mean, T_var, G_mean, G_var), is illustrated in Supplementary Figure S2. Together, these innovations distinguish our LC-JT-VAE from the original JT-VAE by enabling target-specific, conformation-aware molecular glue design.

2.10. Model Training, Optimization, and Implementation Details

The Ligase-Conditioned Junction Tree Variational Autoencoder (LC-JT-VAE) was trained using a β-VAE loss function, consisting of a reconstruction loss and a Kullback–Leibler (KL) divergence term to regularize the latent space [49,50]. The total loss was defined as:

$$\mathcal{L}_{\text{total}} = \mathcal{L}_{\text{recon}} + \beta \cdot \mathcal{L}_{\text{KL}}$$

where $\mathcal{L}_{\text{total}}$ included both molecular graph and junction tree reconstruction losses, and $\mathcal{L}_{\text{KL}}$ promoted latent space continuity. The annealing coefficient β was initialized at 0 and incrementally increased by 0.002 per step until reaching 1.0. Model parameters were initialized with Xavier normal initialization for weights and constant initialization for biases. Training was performed over 2000 epochs using the Adam optimizer (learning rate = 1×10^{-31}), with exponential decay (factor = 0.9 per step) applied through an Exponential LR scheduler. Gradient norms were clipped at a maximum of 50.0 to prevent explosion during backpropagation. Mini-batches were generated using a custom tree-structured molecule loader. For each batch, the model computed the total loss, KL divergence, and three reconstruction metrics: word accuracy (wacc), topology accuracy (tacc), and stereochemistry accuracy (sacc), which were monitored across epochs for convergence evaluation. Model checkpoints were saved at each step for reproducibility and future fine-tuning.

All training, validation, and generation procedures were implemented in Python 3.0 using PyTorch (v1.12) and executed on Google Colab Pro with GPU acceleration [51], utilizing NVIDIA L4 GPUs (22,000+ MiB VRAM), CUDA version 12.4, and driver version 550.54.15.

3. Results

3.1. Identification of a Ligandable Interface on Amyloid-β42 Fibrils

The binding site (BS) analysis revealed three potential binding pockets on the fibrillar assembly, as shown in Figure 3. Of these, two sites were located in clefts (Figure 3a,b) between adjacent protofilaments. While such interfacial regions may appear druggable, their deep burial and steric shielding within the fibril core pose significant challenges for recruiting E3 ligases via bifunctional molecules or molecular glues. These buried sites are unlikely to support the geometric and spatial constraints required for ternary complex formation necessary for targeted protein degradation.

Instead, we selected the third site (Figure 3c), which lies at a more solvent-accessible interfacial region of the fibril surface. This site resides at a junction where structural accessibility and spatial exposure are compatible with simultaneous engagement of a small molecule and an E3 ligase receptor. Given the accessibility of this pocket and its favorable physicochemical enclosure, we prioritized it as the ligandable hotspot for downstream compound docking. Figure 3d illustrates the atomic representation of the binding site residues on Aβ42.

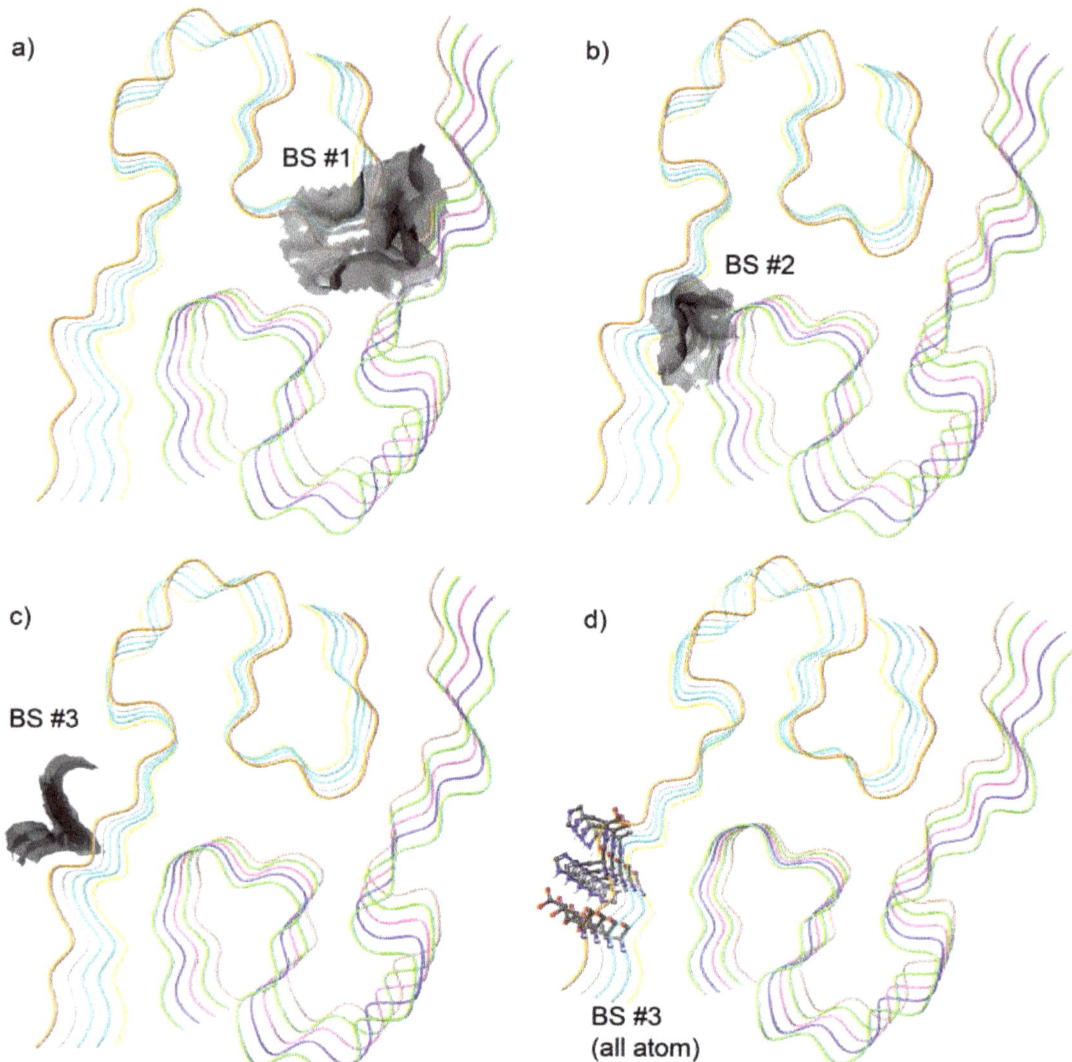

Figure 3. Predicted binding sites (BS) on the amyloid beta 42 (Aβ42) peptide using Schrödinger SiteMap. (**a–c**) Top three ranked potential binding pockets identified on the Aβ42 structure, with surface representations highlighting pocket volumes and spatial locations. Among these, site (**c**) was selected for subsequent molecular docking studies based on druggability and spatial characteristics. (**d**) Atomic-level representation of the selected binding pocket, showing detailed residue arrangement and pocket architecture to guide structure-based ligand design.

3.2. ADMET-Based Filtering of Compound Libraries

To prioritize drug-like compounds suitable for E3 ligase engagement and generative modeling, we applied a series of physicochemical and pharmacokinetic filters to the combined Vitas and ChEMBL compound libraries. The filtering criteria were based on predicted ADMET properties and included molecular weight (MW) between 130 and 725 Da, lipophilicity (logPo/w) between −2 and 6.5, aqueous solubility (logS) between

−6.5 and 0.5, predicted hERG liability (logHERG) < −5, number of predicted metabolic reactions (#metab) between 1 and 8, and no more than one violation of Lipinski's Rule of Five. Following application of these filters, a total of 65,998 compounds were retained across both libraries. These compounds exhibited favorable distributions of drug-like properties, as summarized in Table 1. The average molecular weight was 366.7 Da, with a logP mean of 3.39 and logS mean of −4.59, indicating moderate lipophilicity and acceptable solubility. Most compounds fell within the acceptable range for metabolic stability (mean #metab = 3.64) and had minimal Rule of Five violations (mean = 0.15). This refined set served as the basis for downstream docking, model training, and conditional generation using the JTNNVAE framework.

Table 1. ADMET-based physicochemical profiling of the screened compound derived from Vitas and ChEMBL.

Property	Count	Mean	Std Dev	Min	25%	50%	75%	Max
MW	65,998	366.68	70.31	142.24	316.42	361.35	411.45	668.72
logPo/w	65,998	3.39	1.29	−1.89	2.60	3.54	4.31	6.49
logS	65,998	−4.58	1.29	−6.5	−5.59	−4.77	−3.83	0.47
#metab	65,998	3.64	1.76	1	2	3	5	8
Rule-Of-Five	65,998	0.14	0.35	0	0	0	0	1

3.3. Docking Analysis of Ternary Complexes Formed by E3 Ligases and Amyloid Beta-42 via Molecular Glue Compounds

To investigate the potential of molecular glue compounds in promoting ternary complex formation between Aβ42 and E3 ligases, we performed molecular docking studies involving three key E3 ligases: VHL, CRBN, and MDM2. A representative small molecule with glue-like properties was docked to evaluate its ability to simultaneously interact with both Aβ42 and each of the E3 ligases, thereby stabilizing the formation of a ternary complex. Figure 4a–c panels display the structural configuration of the ternary complexes and the interaction highlights for each E3 ligase. Specifically, the protein structures of VHL, CRBN, and MDM2 are shown in complex with Aβ42 and the docked small molecule. The molecular glue bridges key residues on Aβ42 and the E3 ligases, forming a network of hydrogen bonds, hydrophobic contacts, and π–π stacking interactions essential for ternary complex stabilization.

The 2D interaction diagrams (Figure 4d–f) further delineate the nature of the binding interactions. In the VHL complex, key hydrogen bonds were observed between the molecular glue and residues such as GLU11, HIS13, and TYR112. For the CRBN complex, stabilizing interactions involved TRP86, HIS380, and ASN351. The MDM2 complex exhibited critical interactions with residues including GLU15, HIS19, and TRP23, contributing to the anchoring of the glue compound.

These docking results provide evidence for the potential of specific molecular glues to promote targeted protein degradation of Aβ42 by facilitating the formation of ternary complexes with E3 ligases. Although only one compound is shown here as a representative example, this analysis serves as a proof-of-concept for the broader applicability of molecular glue-based strategies in neurodegenerative disease intervention.

Following docking, molecular dynamics (MDs) simulations were performed on the top-ranked ternary complexes formed by amyloid beta-42, the molecular glue compound, and each E3 ligase (VHL, CRBN, and MDM2). Only the highest-scoring docked compounds were selected for MD analysis to evaluate the stability and dynamic behavior of the ternary complexes over time. These simulations provided additional insights into the conformational stability and key intermolecular interactions sustaining the complex formation. As shown in Supplementary Figure S3, the RMSD trajectories of the VHL–ligand–Aβ42 (S3a), MDM2–ligand–Aβ42 (S3b), and CRBN–ligand–Aβ42 (S3c) complexes remain relatively stable throughout the simulation period, with no significant structural deviations. This suggests that the ternary assemblies are conformationally robust and maintain stable interactions over time. Consistent with this, hydrogen bond analysis further supports the persistence of ligand binding within the ternary interface. Supplementary Figure S4 illustrates the time evolution of hydrogen bonds between the ligand and the E3 ligases in complex with Aβ42. The ligands consistently form multiple hydrogen bonds across the simulation, particularly in the VHL complex (Supplementary Figure S4a), indicating strong and sustained interaction networks. These findings reinforce the potential of our AI-generated compounds to facilitate stable ternary complex formation, a prerequisite for targeted Aβ42 degradation via the ubiquitin–proteasome system.

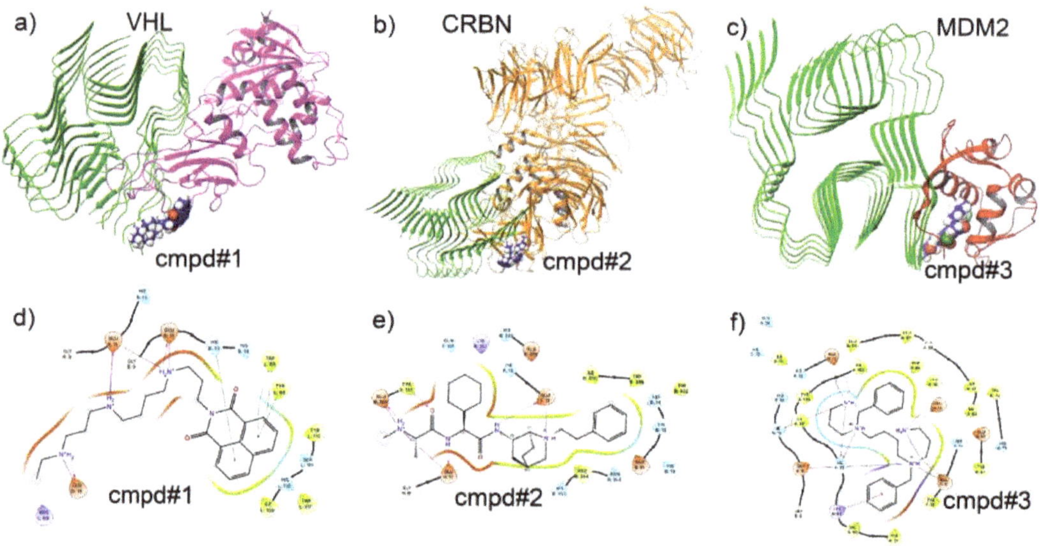

Figure 4. Ternary complex formation between amyloid beta-42 (Aβ42), a representative molecular glue compound, and E3 ligases (VHL, CRBN, and MDM2). (**a–c**) Three-dimensional docking structures showing the formation of ternary complexes for (**a**) VHL–Aβ42–molecular glue, (**b**) CRBN–Aβ42–molecular glue, and (**c**) MDM2–Aβ42–molecular glue. The small molecule compound (cmpd) is shown in ball-and-stick representation, bound at the interface of Aβ42 and each E3 ligase. (**d–f**) Two-dimensional interaction diagrams corresponding to the docking poses for (**d**) VHL, (**e**) CRBN, and (**f**) MDM2 complexes. Intermolecular interactions include hydrogen bonds (purple lines), salt bridges, hydrophobic contacts (green arcs), and π–π stacking, indicating strong binding affinity and ternary complex stabilization.

3.4. Training Data Compounds Characterization to Generate the Ligase-Conditioned Junction Tree Variational Autoencoder (LCJTVAE)

To develop an LCJTVAE model capable of generating novel compounds that facilitate the E3 specific amyloid beta42 degradation, we first classified docking results into three affinity-based categories: High Affinity, Low Affinity, and No Affinity based on docking score thresholds less than −5 kcal/mol, from −5 kcal/mol to −1 kcal/mol and above −1 kcal/mol or positive values, respectively, as shown in Figure 5. Table 2 shows that based on this docking score total of 65,998 compounds from both the library ChEMBL and

Vitas, VHL had the highest number of high-affinity binders (e.g., 8792 from ChEMBL and 6607 from Vitas), while MDM2 exhibited the fewest. This stratification provided a robust and diverse dataset to train our AI model, ensuring that both positive (high/low affinity) and negative (no affinity) examples were well-represented.

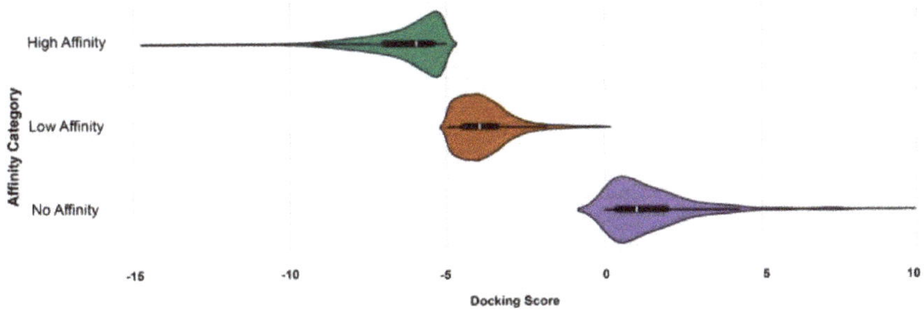

Figure 5. Distribution of docking scores for compounds screened against E3 ligases VHL, CRBN, and MDM2, categorized by affinity class. Docking scores are grouped into three affinity categories: High Affinity (≤−5 kcal/mol), Low Affinity (−5 to −1 kcal/mol), and No Affinity (≥−1 kcal/mol).

Table 2. Summarizes the number of compounds falling into each affinity class for each ligase. A total of 65,998 compounds were evaluated across all conditions.

Ligase	Library	High_Affinity	Low_Affinity	No_Affinity	Total
CRBN	ChEMBL	4834	7448	36	12,318
	Vitas	1748	11,949	73	13,770
MDM2	ChEMBL	131	51	0	182
	Vitas	2302	11,361	86	13,749
VHL	ChEMBL	8792	3414	80	12,286
	Vitas	6607	6949	137	13,693
Total					65,998

We also examined the correlation between docking scores and three key molecular descriptors: molecular weight (MW), lipophilicity (logPo/w), and aqueous solubility (logS) (Figure 6). In Figure 6, high-affinity compounds (blue) were predominantly clustered in the MW range of 300–500 Da, suggesting an optimal size window for ternary complex formation. For logPo/w, most high-affinity compounds fell within the range of 2–5, indicating that moderate lipophilicity may favor binding. In contrast, no-affinity compounds (green) (Figure 6) were more widely distributed and skewed toward higher solubility values (logS > −2), whereas high-affinity compounds generally had lower logS values (<−4), reflecting reduced aqueous solubility. These quantitative trends highlight that docking performance and binding potential are influenced by distinct ADMET property windows, which were integrated as features in the training dataset for the AI model.

To further dissect the structural characteristics associated with varying binding affinities, we conducted a scaffold analysis of compounds across the high, low, and no-affinity categories for each E3 ligase (VHL, CRBN, and MDM2) (Figure 7). Distinct trends emerged for each ligase, suggesting that specific structural motifs are favored depending on the binding partner.

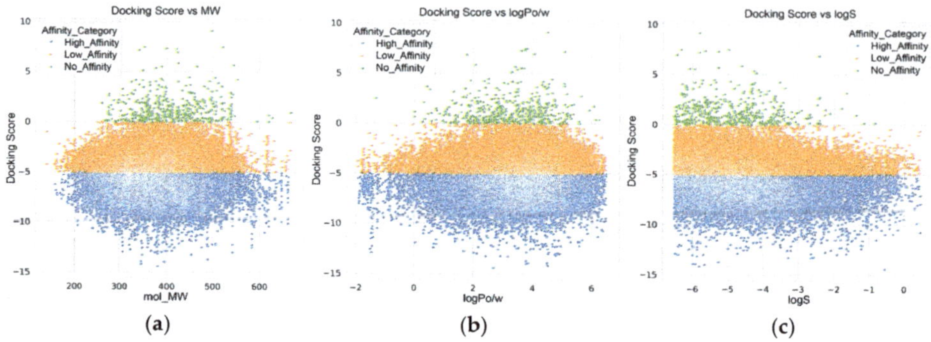

Figure 6. Relationship between docking scores and selected ADMET properties across all screened compounds. Compounds are color-coded by affinity category: High Affinity (blue), Low Affinity (orange), and No Affinity (green). (**a**) Docking scores vs. molecular weight (MW); (**b**) docking scores vs. predicted aqueous solubility (logS); (**c**) docking scores vs. predicted lipophilicity (logP). The plots illustrate the spread of affinity categories across diverse physicochemical properties, revealing that high-affinity compounds tend to cluster within drug-like regions, supporting their potential suitability for further development.

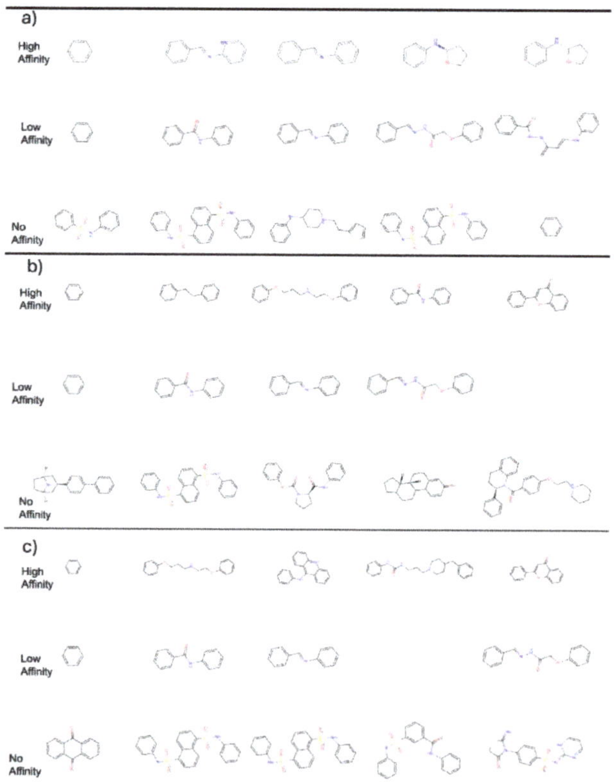

Figure 7. Most frequently occurring molecular scaffolds identified across three affinity categories—High Affinity (top), Low Affinity (middle), and No Affinity (bottom)—for each E3 ligase: (**a**) VHL, (**b**) MDM2, and (**c**) CRBN.

For VHL (Figure 7a), high-affinity compounds predominantly contained substituted aromatic rings and flexible linkers incorporating nitrogen atoms, while low-affinity compounds were structurally simpler, often lacking key heteroatoms. No-affinity compounds tended to feature bulky fused-ring systems and sulfur-containing groups, which may contribute to steric hindrance or poor complementarity at the binding site.

In the case of MDM2 (Figure 7b), high-affinity scaffolds were characterized by biphenyl motifs and polar side chains, likely enhancing both hydrophobic packing and hydrogen bonding. In contrast, no-affinity scaffolds were bulkier and more rigid, often featuring heterocycles and extended conjugation.

For CRBN (Figure 7c), high-affinity compounds commonly included ether-linked alkyl chains and small heterocycles, suggesting that flexibility and size may be important determinants of binding. No-affinity compounds frequently exhibited complex fused polyaromatic systems and sulfur-containing linkers, which may disrupt favorable binding orientations. This scaffold-based categorization not only validates the docking-based affinity predictions but also highlights recurring ligase-specific chemical features, which are essential for guiding structure-aware AI model training and scaffold-based virtual screening efforts.

3.5. Graph and Junction Tree Representations of Molecules for Model Training

To facilitate the generation of chemically valid and interpretable molecular structures, we utilized dual representations for each molecule: a molecular graph and a junction tree, serving as structured inputs to the Ligase-Conditioned JT-VAE framework.

Each molecule was decomposed into a set of cliques—maximal subgraphs such as aromatic rings or functional groups—to capture its structural hierarchy (Supplementary Figure S5a). For example, in one representative compound, a brominated aromatic ring (black-circled) and a hydroxyl-substituted phenyl group (red-circled) were identified as distinct cliques. These cliques were assembled into a junction tree structure, where each node represents a clique and edges represent direct atom-sharing connections between cliques (Supplementary Figure S5b). In this example, node 14 corresponds to the brominated ring and node 15 to the hydroxyl-phenyl group, with node indices matching the entries in Supplementary Table S1, which lists the atom indices defining each clique.

During molecule generation, the model first predicts the junction tree scaffold to ensure chemical validity, followed by assembly of the full molecular graph based on atomic connectivity between cliques. The molecular graph representation (Supplementary Figure S5c) captures atomic-level detail, with nodes representing atoms and edges representing bonds; standard bonds are shown in blue, while bonds participating in rings are highlighted in red. This detailed graph is used in the assembly phase to accurately reconstruct chemically coherent structures. A complete listing of all cliques and corresponding edges for the training compounds (up to 50 molecules) is provided in Supplementary Table S2.

3.6. The Loss Function

The ligase-conditioned JTNNVAE model was trained for 500 epochs using the Adam optimizer with an initial learning rate of 0.001 and an exponential learning rate decay (decay rate = 0.9 per step). To stabilize the training dynamics, all bias parameters were initialized to zero, and non-bias parameters were initialized using Xavier normal initialization. A KL annealing schedule was applied during training, with the KL divergence coefficient (β) linearly increased from 0 to 1 at a rate of 0.002 per batch.

To prevent gradient explosion, gradient clipping was applied with a max norm of 50. The total number of model parameters was approximately 5393 K. The training loss, plotted in Figure 8, shows a sharp decline within the first 100–150 epochs, followed by a gradual convergence toward lower values, indicating stable model optimization.

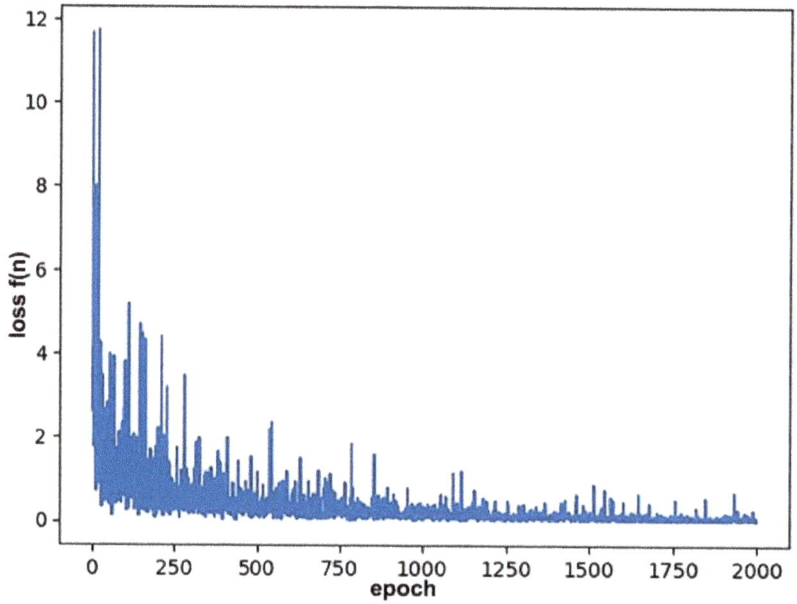

Figure 8. Training loss curve of the JTNNVAE model trained on docked compounds targeting Amyloid beta42. The sharp initial drop followed by gradual convergence indicates successful learning of chemically valid structures relevant to E3 ligase binding.

3.7. *AI-Generated Compounds and Model Performance*

3.7.1. Molecular Diversity Visualization via t-SNE

To assess the chemical diversity and distribution of AI-generated compounds relative to training molecules, we employed t-distributed stochastic neighbor embedding (t-SNE) to project high-dimensional molecular fingerprints into two-dimensional space. To enable clear visualization of the chemical space, we randomly selected 100 training compounds for each E3 ligase (VHL, CRBN, and MDM2). Given that the number of AI-generated compounds was relatively small (up to 10 maximum), using the full training dataset—which is much larger—would have obscured the spatial distribution and diversity of the generated molecules. Figure 9 displays t-SNE plots for VHL, MDM2, and CRBN-specific compound sets, visualizing both training compounds (blue) and AI-generated molecules (orange) in their respective embedding spaces.

For VHL (Figure 9a), the AI-generated compounds clustered tightly in a distinct region of the chemical space, yet remained within the broader distribution of the training molecules. This suggests that the model effectively captured scaffold-level diversity while preserving chemical validity. In the case of MDM2 (Figure 9b), generated compounds exhibited partial overlap with the training distribution, indicating moderate exploration of novel chemical space while still retaining features relevant to the target ligase. For CRBN (Figure 9c), the generated compounds formed a dense and centrally located cluster, reflecting the model's focus on optimizing around known chemical space with a bias toward validity and synthesizability.

Collectively, these visualizations demonstrate that the Ligase-Conditioned JT-VAE framework is capable of generating structurally diverse and target-specific molecular glues, while modulating novelty depending on training data distribution and target binding site complexity.

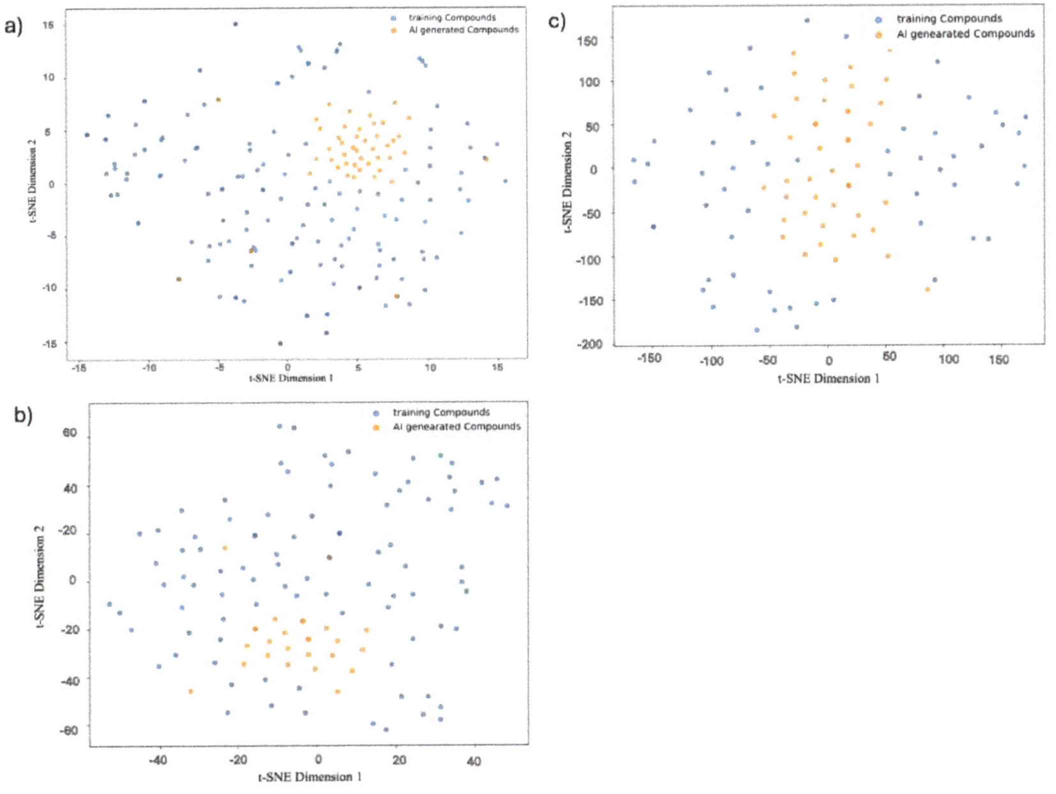

Figure 9. t-SNE plots visualizing the chemical space distribution of training compounds (blue) and AI-generated compounds (orange) for each E3 ligase: (**a**) VHL, (**b**) MDM2, and (**c**) CRBN.

3.7.2. Model Performance Across E3 Ligases

To evaluate the quality of molecules generated by our Ligase-Conditioned JT-VAE (LC-JT-VAE) framework, we assessed five key performance metrics: validity, uniqueness, novelty, average quantitative estimate of drug-likeness (QED), and Lipinski's rule-of-five adherence across three target E3 ligases: VHL, MDM2, and CRBN (Figure 10). All models consistently produced highly valid molecules, with validity scores exceeding 96%, confirming the chemical feasibility of the generated SMILES strings. The VHL model achieved the highest performance, with a validity of 96.30%, uniqueness of 85.19%, and novelty of 81.48%, indicating strong diversity and minimal overlap with the training set.

The CRBN-specific model generated molecules with 100% validity, although with slightly lower uniqueness (82%) and novelty (80%), suggesting a higher overlap with learned structural motifs. The MDM2 model showed balanced performance with 97.87% validity, 97.87% uniqueness, and 93.62% novelty, demonstrating strong generalization beyond training data. However, QED scores across all models remained moderate (0.35–0.42), reflecting the potential need for post hoc optimization to enhance drug-likeness. Similarly,

Lipinski's adherence ranged from 60 to 83%, with the VHL-targeted model outperforming CRBN and MDM2, suggesting a better balance between molecular complexity and pharmacokinetic profiles. These metrics validate the LC-JT-VAE model's capability to generate novel, chemically valid, and E3-selective molecular glues, with each E3-specific model learning distinct chemical priors reflective of its ligase binding site context.

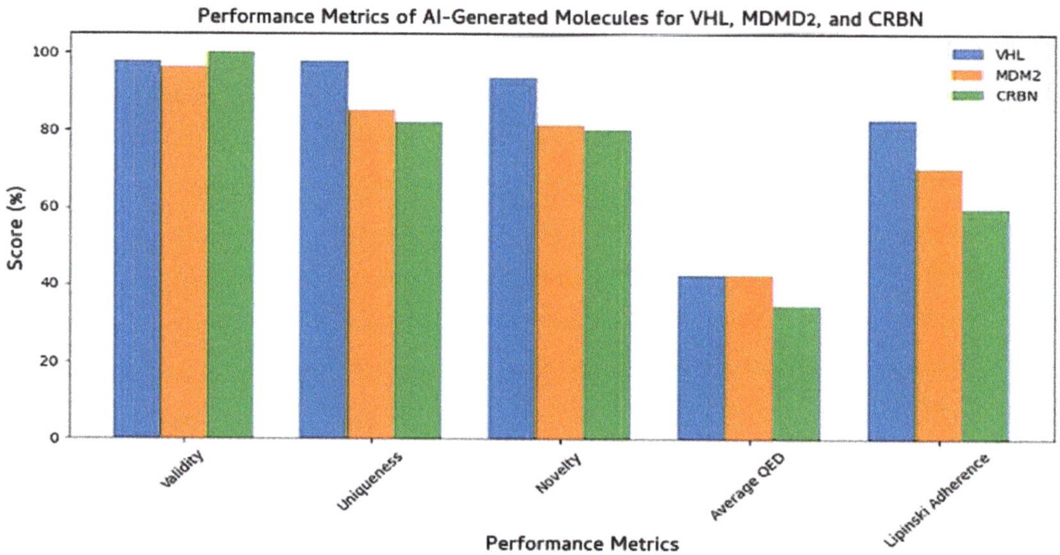

Figure 10. Performance evaluation of AI-generated molecules for E3 ligases VHL, MDM2, and CRBN across key generative metrics. Metrics include validity, uniqueness, and novelty of generated compounds, as well as average quantitative estimate of drug-likeness (QED) and compliance with Lipinski's Rule of Five. High validity and novelty scores indicate the model's ability to generate chemically sound and diverse structures, while QED and Lipinski adherence reflect drug-like potential of the designed molecules.

3.7.3. E3 Ligase Protein-Specific Structural Insights into AI-Generated Compounds

The generated compounds for VHL (Figure 11a) displayed diverse ring systems, aliphatic chains, and heteroatoms (e.g., nitrogen, oxygen) that are consistent with features found in known VHL binders. Notably, several molecules contained ester, amide, and cyclic amine functionalities, which are important pharmacophoric elements for engaging VHL binding pockets.

For MDM2 (Figure 11b), the model generated molecules having scaffolds enriched with aromatic and polycyclic structures, some bearing fused heterocycles or alkyl substitutions that resemble known MDM2-p53 interaction inhibitors. The generated compounds often featured extended planar regions, suggesting favorable π–π stacking potential within the MDM2 binding cleft.

In the case of CRBN (Figure 11c), the generated compounds included many small fragment-like structures and thalidomide-inspired motifs such as phthalimide cores and flexible linkers. These molecules tended to be chemically simpler yet retained functional groups commonly associated with CRBN modulators, indicating the model's ability to learn and replicate CRBN-specific ligand patterns.

Together, these AI-generated molecular glues show ligand diversity tailored to the unique binding environments of VHL, MDM2, and CRBN, supporting the effectiveness of our protein-conditioned generation strategy.

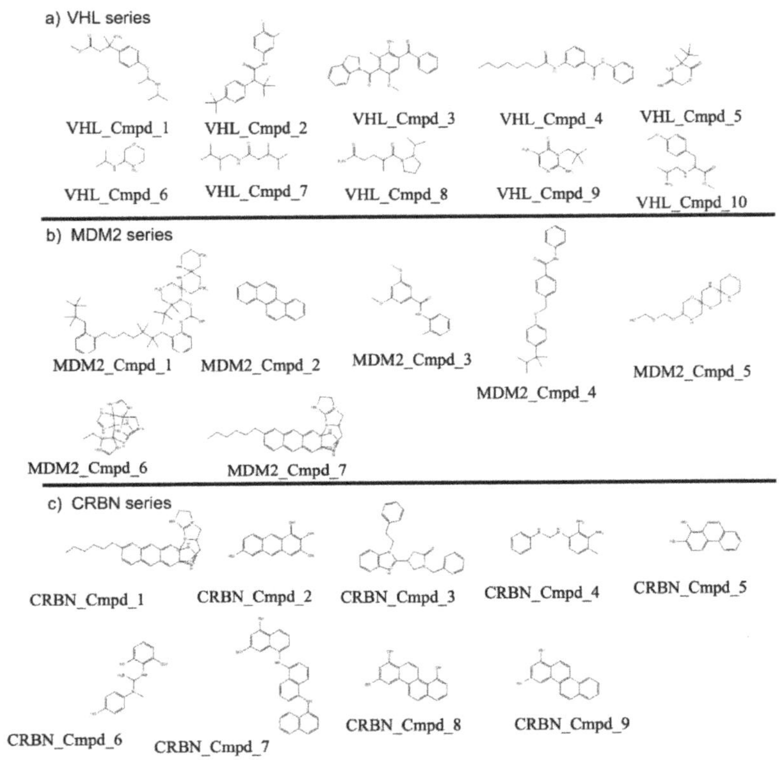

Figure 11. Representative AI-generated compounds designed for three E3 ligases: (**a**) VHL, (**b**) MDM2, and (**c**) CRBN. The series of exemplar compounds illustrates the distinct scaffolding preferences and functional group diversity observed from our results.

3.7.4. Three-Dimensional Conformational Plausibility of Ligase-Specific Compounds

To assess the 3D conformational stability of the ligase-specific compounds generated by our model, we applied RDKit's ETKDG algorithm to independently generate multiple conformers for each molecule. Although the Supplementary Tables S3–S5 do not report absolute energy values, the use of RDKit's ETKDG algorithm—which is designed to generate low-energy, experimentally plausible conformations—combined with the consistently low RMSD values between independent conformers, supports the interpretation that the generated ligase-specific compounds adopt stable and chemically reasonable 3D geometries. The conformational robustness of each compound was evaluated by calculating the root-mean-square deviation (RMSD) between two independently generated conformers (Supplementary Tables S3–S5). RMSD values were negligibly small across all compounds—ranging from 0 to approximately 1.3×10^{-7} Å—even for highly flexible molecules such as VHL_Cmpd_1 (13 rotatable bonds), VHL_Cmpd_8 (12 rotatable bonds) (Supplementary Table S3), and MDM2_Cmpd_6 (22 rotatable bonds) (Supplementary Table S5). These minimal deviations indicate that the generated molecules consistently adopt low-energy, chemically plausible conformations, further validating the inclusion of torsional angle features in our Ligase-Conditioned JT-VAE framework. This analysis supports the model's capacity to produce structurally reliable compounds suitable for downstream structure-based drug design. The Python script used for this evaluation is provided in the Supplementary Section (Figure S7).

3.8. Target-Specific Docking Validates the Precision of AI-Generated Compounds

To evaluate the binding specificity and interaction profiles of our AI-generated compounds, we performed systematic molecular docking of all designed molecules against each of the three E3 ligase–Aβ42 complexes (VHL, CRBN, and MDM2). For each compound, docking was carried out not only against its intended E3 ligase target but also cross-docked into the other two ligases to assess potential off-target binding. Supplementary Section Figure S6 presents examples of the top-ranked compounds based on docking scores for each E3 ligase: VHL_Cmpd_4 for VHL (docking score: −5.98 kcal/mol in Figure 12), CRBN_Cmpd_3 for CRBN (−5.80 kcal/mol in Figure 12), and MDM2_Cmpd_5 for MDM2 (−5.78 kcal/mol in Figure 12). Structural visualizations in Figure 12 illustrate that the topmost-ranked docked compounds engage their respective E3 ligase binding interfaces through multiple key interactions, including hydrogen bonding, hydrophobic contacts, and π–π stacking with critical pocket residues. For instance, VHL_Cmpd_4 forms hydrogen bonds with HIS13 of Aβ42 and HIS110 of VHL, whereas CRBN_Cmpd_2 establishes extensive interactions with HIS355, TRP402, and surrounding residues within the CRBN binding pocket.

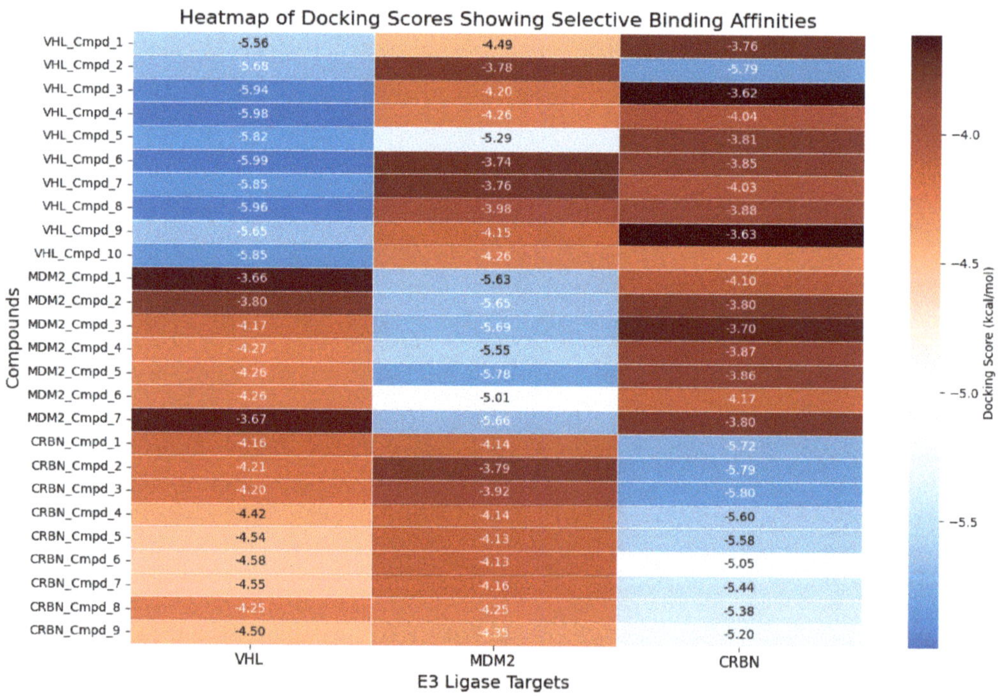

Figure 12. Heatmap of docking scores (kcal/mol) for AI-generated compounds across VHL, CRBN, and MDM2 ligases. Each row represents a compound, and each column corresponds to an E3 ligase. Stronger (more negative) binding affinities are shown in blue. The results demonstrate target-specific binding, with designed compounds showing preferential affinity for their intended ligases and lesser binding affinity for their unintended targets.

To quantify the binding selectivity of all generated compounds, a heatmap of docking scores across all ligand–ligase combinations is shown in Figure 12. Compounds generally demonstrated stronger binding (more negative docking scores) to their designed target ligase compared to off-target ligases. Specifically, the average docking score for VHL-specific compounds against VHL was −5.84 kcal/mol, compared to an average of −4.05 kcal/mol and −4.15 kcal/mol when docked into MDM2 and CRBN, respectively. Similarly, CRBN-specific compounds exhibited an average docking score of −5.76 kcal/mol against CRBN, whereas the corresponding averages against VHL and MDM2 were markedly less favorable. These quantitative results further confirm that our LC-JT-VAE platform successfully generated compounds with strong binding affinity and target selectivity toward their intended E3 ligase.

4. Discussion

In this study, we developed a structure-guided generative framework to design E3 ligase-specific molecular glues targeting amyloid-β42 (Aβ42) fibrils. Starting with the identification of a ligandable site on the Aβ42 fibril surface (Figure 3), we screened a large chemical space to discover compounds capable of bridging Aβ42 with key E3 ligases—CRBN, VHL, and MDM2. Following rigorous physicochemical and pharmacokinetic filtering, candidate molecules were evaluated for their ability to form ternary complexes through interaction modeling with each E3 ligase and Aβ42 (Table 2). These prioritized compounds were subsequently used to train a target-specific generative model that integrates both the structural features of ligase binding sites and the 3D conformational dynamics of small molecules, incorporating torsional angles as an additional feature to enhance chemical realism.

Our results demonstrate that the generative model effectively captured ligase-specific chemical preferences, producing valid, novel, and structurally diverse compounds optimized for each E3 ligase (Figures 9 and 10). Performance evaluation showed consistently high chemical validity and target selectivity across generated compound sets (Figures 10 and 12). Dimensionality reduction analysis confirmed that the model generated molecules covering both novel and known chemical spaces, while detailed compound visualizations revealed diversity and preservation of critical pharmacophores associated with ligase engagement (Figure 11). These outcomes highlight the model's potential utility for designing degrader compounds against structurally challenging targets such as Aβ42.

The comprehensive docking analysis further underscores the effectiveness of our LC-JT-VAE generative framework in designing compounds with targeted binding specificity. By docking all generated compounds across multiple E3 ligases, we demonstrated that the molecules preferentially bind to their intended ligase targets rather than interacting nonspecifically across ligases. This specificity is crucial for the success of molecular glue–mediated ternary complex formation.

The top-ranked compounds, presented in Figure S6, provide illustrative examples of how our designed molecules establish favorable interaction networks within the ligase binding pockets, engaging critical residues through hydrogen bonding, hydrophobic contacts, and π–π interactions—key mechanisms for stabilizing ternary complexes.

Moreover, the cross-target docking heatmap (Figure 12) quantitatively supports this finding, showing that compounds exhibit significantly stronger binding (i.e., more negative docking scores) to their respective intended E3 ligases compared to off-target ligases. These results collectively validate the robustness of our AI-driven design platform, confirming its ability to generate chemically valid, structurally optimized, and functionally targeted therapeutic candidates.

Together, these findings position our integrated predictive and generative pipeline as a powerful strategy for early intervention in age-related diseases such as Alzheimer's disease, and offer a generalizable framework for therapeutic discovery against other un-druggable targets.

5. Conclusions

This study presents an AI-driven framework for the rational design of E3 ligase–specific molecular glues targeting Aβ42, a key pathogenic peptide in Alzheimer's disease. By integrating ligand- and structure-based screening with advanced generative modeling, we demonstrate that the Ligase-Conditioned JT-VAE (LC-JT-VAE) can generate novel, drug-like compounds tailored for CRBN, VHL, and MDM2 ligases. These compounds exhibit strong potential to facilitate ternary complex formation and promote targeted degradation of intracellular Aβ42. The integration of protein sequence embeddings and torsional flexibility into the generative model significantly improves target specificity and chemical realism. Our approach sets the stage for AI-enabled discovery of therapeutic degraders against aggregation-prone proteins, offering new avenues in the treatment of Alzheimer's and other neurodegenerative disorders.

Supplementary Materials: The following supporting information can be downloaded at: https://www.mdpi.com/article/10.3390/biom15060849/s1, Figure S1. Dihedral angles computed using RDKit's GetDihedralDeg function. (a) RDKit code snippet illustrating hydrogen addition (AddHs), molecular geometry optimization using the MMFF94 force field (MMFFOptimizeMolecule), and retrieval of the optimized conformer (GetConformer) for a given molecule. (b) Custom Python function get_torsional_angle for computing the torsional (dihedral) angle across a rotatable bond. The function identifies neighboring atoms required for angle computation and uses *Chem.rdMolTransforms.GetDihedralDeg* to calculate the dihedral angle in degrees based on the conformer coordinates. Figure S2. Architecture of the JTNNVAE model for ligase-conditioned molecular glue generation. The model consists of a junction tree encoder (JTNNEncoder) that embeds molecular graphs into a latent space, a graph-based message-passing network (GRUGraphGRU) to capture atom-level connectivity, and a junction tree decoder (JTNNDecoder) for reconstructing chemically valid clique structures. An additional binding site residue encoder processes E3 ligase binding site features (e.g., residue embeddings from ProtBERT), enabling the model to condition generation on specific ligase contexts. The MPN encoder integrates atomic features, while the atom assembly module combines latent vectors from the junction tree and residue encoder to assemble complete molecules. Finally, the model outputs ligase-specific compounds via three linear transformations (T_real, T_mean, and G_var) that capture the conditional generation dynamics. This architecture enables the targeted design of molecular glue compounds tailored to specific E3 ligase binding environments. Figure S3. Root-mean-square deviation (RMSD) plots for ternary complexes of Aβ42, E3 ligases, and top-ranked ligands over 200 ns molecular dynamics simulations. (a) VHL–ligand–Aβ42 complex, (b) MDM2–ligand–Aβ42 complex, and (c) CRBN–ligand–Aβ42 complex. RMSD values (in A) were calculated for the entire ternary complex to evaluate structural stability over time. The relatively stable trajectories across all panels indicate that the ternary assemblies remained conformationally stable throughout the simulation period, supporting the feasibility of molecular glue–mediated complex formation. Figure S4. Time evolution of hydrogen bond interactions between the top-scoring ligand and its corresponding E3 ligase–Aβ42 complex over a 100 ns molecular dynamics simulation. (a) VHL-ligand complex, (b) MDM2-ligand complex, and (c) CRBN-ligand complex. The y-axis indicates the number of hydrogen bonds formed between the ligand and protein throughout the simulation, while the x-axis shows the simulation time in nanoseconds (ns). These interaction profiles provide insights into the stability and persistence of ligand binding within the ternary complex interface. Figure S5. Graph and Junction Tree representations of a molecule used for JTNN-VAE training. (a) Original molecular structure annotated with atom indices. Two representative substructures (cliques) are highlighted: a brominated aromatic ring (black circle) and a hydroxylsubstituted phenyl ring (red circle). (b) Junction tree representation of the molecule, where each node corresponds to a chemically valid clique (e.g., ring, functional group, or bridged system). The circled nodes (14 and 15) represent the cliques highlighted in the molecular structure. Edges represent connectivity based on shared atoms between cliques. (c) Molecular graph representation, where atoms are nodes and bonds are edges. Blue edges represent standard bonds, while red edges denote bonds participating in ring structures. Figure S6. Top most ranked AI-generated compounds docked to E3 ligase–Aβ42 complexes. (a) VHL_Cmpd_4 bound to VHL–Aβ42 complex; (b) CRBN_Cmpd_2 bound to CRBN–Aβ42 complex; (c) MDM2_Cmpd_5 bound to MDM2–Aβ42 complex. (d–f) Corresponding 2D interaction diagrams showing key hydrogen bonds and hydrophobic contacts stabilizing each ligand within the E3 binding sites. Figure S7. Python script used to evaluate the 3D conformational plausibility of ligase-specific compounds. Table S1. List of cliques used in the junction tree, with each bracket representing a set of atom indices forming a maximal subgraph (clique). Indices match the atom numbers from the original molecule and the junction tree nodes. Table S2. The 50 exemplar molecular compounds showing the SMILES string representation along with its corresponding junction tree structure, which includes cliques (clusters of atoms forming subgraphs) and edges (connections between cliques). Table S3. Assessment of 3D conformational plausibility of VHL-targeting compounds generated by the Ligase-Conditioned JT-VAE model. Table S4. Assessment of 3D conformational plausibility of CRBN-targeting compounds generated by the Ligase-Conditioned JT-VAE model. Table S5. Assessment of 3D conformational plausibility of MDM2-targeting compounds generated by the Ligase-Conditioned JT-VAE model.

Author Contributions: Conceptualization, N.N.I. and T.R.C.; methodology, N.N.I. and T.R.C.; software, N.N.I. and T.R.C.; validation, N.N.I. and T.R.C.; formal analysis, N.N.I. and T.R.C.; investigation, N.N.I. and T.R.C.; resources, N.N.I. and T.R.C.; data curation, N.N.I. and T.R.C.; writing, N.N.I. and T.R.C.; visualization, N.N.I.; supervision, T.R.C.; project administration, T.R.C.; funding acquisition, T.R.C. All authors have read and agreed to the published version of the manuscript.

Funding: T.R.C. was supported in part by the Center for Individualized Medicine (CIM) and the Florida Department of Health—Ed and Ethel Moore Alzheimer's Disease Research Program [24A19].

Institutional Review Board Statement: Not applicable.

Informed Consent Statement: Not applicable.

Data Availability Statement: The complete dataset and all relevant code used in this study are publicly available in the GitHub repository accessible at: https://github.com/thom-DEC/LCJTVAE-SupplementalData.

Conflicts of Interest: N.N.I and T.R.C. declare they have no competing interests. The funders had no role in the design of the study; in the collection, analyses, or interpretation of data; in the writing of the manuscript; or in the decision to publish the results.

References

1. Selkoe, D.J.; Hardy, J. The amyloid hypothesis of Alzheimer's disease at 25 years. *EMBO Mol. Med.* **2016**, *8*, 595–608. [CrossRef]
2. Hoe, H.S.; Lee, H.K.; Pak, D.T. The upside of APP at synapses. *CNS Neurosci. Ther.* **2012**, *18*, 47–56. [CrossRef] [PubMed]
3. Haass, C.; Selkoe, D.J. Soluble protein oligomers in neurodegeneration: Lessons from the Alzheimer's amyloid beta-peptide. *Nat. Rev. Mol. Cell Biol.* **2007**, *8*, 101–112. [CrossRef] [PubMed]
4. Koch, L. Non-coding RNA: A protective role for TERRA at telomeres. *Nat. Rev. Genet.* **2017**, *18*, 453. [CrossRef]
5. Li, Y.; Schindler, S.E.; Bollinger, J.G.; Ovod, V.; Mawuenyega, K.G.; Weiner, M.W.; Shaw, L.M.; Masters, C.L.; Fowler, C.J.; Trojanowski, J.Q.; et al. Validation of Plasma Amyloid-beta 42/40 for Detecting Alzheimer Disease Amyloid Plaques. *Neurology* **2022**, *98*, e688–e699. [CrossRef]
6. Hampel, H.; Hardy, J.; Blennow, K.; Chen, C.; Perry, G.; Kim, S.H.; Villemagne, V.L.; Aisen, P.; Vendruscolo, M.; Iwatsubo, T.; et al. The Amyloid-β Pathway in Alzheimer's Disease. *Mol. Psychiatry* **2021**, *26*, 5481–5503. [CrossRef] [PubMed]

7. Lu, M.; Williamson, N.; Mishra, A.; Michel, C.H.; Kaminski, C.F.; Tunnacliffe, A.; Kaminski Schierle, G.S. Structural progression of amyloid-beta Arctic mutant aggregation in cells revealed by multiparametric imaging. *J. Biol. Chem.* **2019**, *294*, 1478–1487. [CrossRef]
8. Lin, J.Y.; Wang, W.A.; Zhang, X.; Liu, H.Y.; Zhao, X.L.; Huang, F.D. Intraneuronal accumulation of Abeta42 induces age-dependent slowing of neuronal transmission in Drosophila. *Neurosci. Bull.* **2014**, *30*, 185–190. [CrossRef]
9. Zeidel, M.L.; Hammond, T.; Botelho, B.; Harris, H.W., Jr. Functional and structural characterization of endosomes from toad bladder epithelial cells. *Am. J. Physiol.* **1992**, *263*, F62–F76. [CrossRef]
10. Hardy, J.A.; Higgins, G.A. Alzheimer's disease: The amyloid cascade hypothesis. *Science* **1992**, *256*, 184–185. [CrossRef]
11. Imbimbo, B.P.; Panza, F.; Frisardi, V.; Solfrizzi, V.; D'Onofrio, G.; Logroscino, G.; Seripa, D.; Pilotto, A. Therapeutic intervention for Alzheimer's disease with gamma-secretase inhibitors: Still a viable option? *Expert. Opin. Investig. Drugs* **2011**, *20*, 325–341. [CrossRef] [PubMed]
12. Sevigny, J.; Suhy, J.; Chiao, P.; Chen, T.; Klein, G.; Purcell, D.; Oh, J.; Verma, A.; Sampat, M.; Barakos, J. Amyloid PET Screening for Enrichment of Early-Stage Alzheimer Disease Clinical Trials: Experience in a Phase 1b Clinical Trial. *Alzheimer Dis. Assoc. Disord.* **2016**, *30*, 1–7. [CrossRef]
13. van Dyck, C.H.; Swanson, C.J.; Aisen, P.; Bateman, R.J.; Chen, C.; Gee, M.; Kanekiyo, M.; Li, D.; Reyderman, L.; Cohen, S.; et al. Lecanemab in Early Alzheimer's Disease. *N. Engl. J. Med.* **2023**, *388*, 9–21. [CrossRef]
14. Menzies, F.M.; Fleming, A.; Caricasole, A.; Bento, C.F.; Andrews, S.P.; Ashkenazi, A.; Fullgrabe, J.; Jackson, A.; Jimenez Sanchez, M.; Karabiyik, C.; et al. Autophagy and Neurodegeneration: Pathogenic Mechanisms and Therapeutic Opportunities. *Neuron* **2017**, *93*, 1015–1034. [CrossRef] [PubMed]
15. Elsworthy, R.J.; Crowe, J.A.; King, M.C.; Dunleavy, C.; Fisher, E.; Ludlam, A.; Parri, H.R.; Hill, E.J.; Aldred, S. The effect of citalopram treatment on amyloid-beta precursor protein processing and oxidative stress in human hNSC-derived neurons. *Transl. Psychiatry* **2022**, *12*, 285. [CrossRef]
16. Chung, C.W.; Stephens, A.D.; Konno, T.; Ward, E.; Avezov, E.; Kaminski, C.F.; Hassanali, A.A.; Kaminski Schierle, G.S. Intracellular Abeta42 Aggregation Leads to Cellular Thermogenesis. *J. Am. Chem. Soc.* **2022**, *144*, 10034–10041. [CrossRef] [PubMed]
17. Ohyagi, Y.; Tsuruta, Y.; Motomura, K.; Miyoshi, K.; Kikuchi, H.; Iwaki, T.; Taniwaki, T.; Kira, J. Intraneuronal amyloid beta42 enhanced by heating but counteracted by formic acid. *J. Neurosci. Methods* **2007**, *159*, 134–138. [CrossRef]
18. Kitajima, Y.; Yoshioka, K.; Suzuki, N. The ubiquitin–proteasome system in regulation of the skeletal muscle homeostasis and atrophy: From basic science to disorders. *J. Physiol. Sci.* **2020**, *70*, 40. [CrossRef]
19. Zhang, Y.; Chen, H.; Li, R.; Sterling, K.; Song, W. Amyloid β-based therapy for Alzheimer's disease: Challenges, successes and future. *Signal Transduct. Target. Ther.* **2023**, *8*, 248. [CrossRef]
20. Wertz, I.E.; Wang, X. From Discovery to Bedside: Targeting the Ubiquitin System. *Cell Chem. Biol.* **2019**, *26*, 156–177. [CrossRef]
21. Thibaudeau, T.A.; Anderson, R.T.; Smith, D.M. A common mechanism of proteasome impairment by neurodegenerative disease-associated oligomers. *Nat. Commun.* **2018**, *9*, 1097. [CrossRef] [PubMed]
22. King, E.A.; Cho, Y.; Hsu, N.S.; Dovala, D.; McKenna, J.M.; Tallarico, J.A.; Schirle, M.; Nomura, D.K. Chemoproteomics-enabled discovery of a covalent molecular glue degrader targeting NF-kappaB. *Cell Chem. Biol.* **2023**, *30*, 394–402.e9. [CrossRef] [PubMed]
23. Hanzl, A.; Winter, G.E. Targeted protein degradation: Current and future challenges. *Curr. Opin. Chem. Biol.* **2020**, *56*, 35–41. [CrossRef] [PubMed]
24. Potjewyd, F.M.; Axtman, A.D. Exploration of Aberrant E3 Ligases Implicated in Alzheimer's Disease and Development of Chemical Tools to Modulate Their Function. *Front. Cell. Neurosci.* **2021**, *15*, 768655. [CrossRef]
25. Ivanenkov, Y.A.; Polykovskiy, D.; Bezrukov, D.; Zagribelnyy, B.; Aladinskiy, V.; Kamya, P.; Aliper, A.; Ren, F.; Zhavoronkov, A. Chemistry42: An AI-Driven Platform for Molecular Design and Optimization. *J. Chem. Inf. Model.* **2023**, *63*, 695–701. [CrossRef]
26. McNair, D. Artificial Intelligence and Machine Learning for Lead-to-Candidate Decision-Making and Beyond. *Annu. Rev. Pharmacol. Toxicol.* **2023**, *63*, 77–97. [CrossRef]
27. Singh, S.; Gupta, H.; Sharma, P.; Sahi, S. Advances in Artificial Intelligence (AI)-assisted approaches in drug screening. *Artif. Intell. Chem.* **2024**, *2*, 100039. [CrossRef]
28. Paul, D.; Sanap, G.; Shenoy, S.; Kalyane, D.; Kalia, K.; Tekade, R.K. Artificial intelligence in drug discovery and development. *Drug Discov. Today* **2021**, *26*, 80–93. [CrossRef]
29. Perron, Q.; Mirguet, O.; Tajmouati, H.; Skiredj, A.; Rojas, A.; Gohier, A.; Ducrot, P.; Bourguignon, M.P.; Sansilvestri-Morel, P.; Do Huu, N.; et al. Deep generative models for ligand-based de novo design applied to multi-parametric optimization. *J. Comput. Chem.* **2022**, *43*, 692–703. [CrossRef]
30. Luukkonen, S.; van den Maagdenberg, H.W.; Emmerich, M.T.M.; van Westen, G.J.P. Artificial intelligence in multi-objective drug design. *Curr. Opin. Struct. Biol.* **2023**, *79*, 102537. [CrossRef]

31. Arus-Pous, J.; Johansson, S.V.; Prykhodko, O.; Bjerrum, E.J.; Tyrchan, C.; Reymond, J.L.; Chen, H.; Engkvist, O. Randomized SMILES strings improve the quality of molecular generative models. *J. Cheminform.* **2019**, *11*, 71. [CrossRef]
32. Reiser, P.; Neubert, M.; Eberhard, A.; Torresi, L.; Zhou, C.; Shao, C.; Metni, H.; van Hoesel, C.; Schopmans, H.; Sommer, T.; et al. Graph neural networks for materials science and chemistry. *Commun. Mater.* **2022**, *3*, 93. [CrossRef]
33. Meyers, J.; Fabian, B.; Brown, N. De novo molecular design and generative models. *Drug Discov. Today* **2021**, *26*, 2707–2715. [CrossRef]
34. Ramirez-Mora, T.; Retana-Lobo, C.; Valle-Bourrouet, G. Biochemical characterization of extracellular polymeric substances from endodontic biofilms. *PLoS ONE* **2018**, *13*, e0204081. [CrossRef]
35. Jin, W.; Barzilay, R.; Jaakkola, T. Junction Tree Variational Autoencoder for Molecular Graph Generation. *arXiv* **2019**, arXiv:1802.04364.
36. Truong, G.-B.; Pham, T.-A.; To, V.-T.; Le, H.-S.L.; Van Nguyen, P.-C.; Trinh, T.-C.; Phan, T.-L.; Truong, T.N. Discovery of Vascular Endothelial Growth Factor Receptor 2 Inhibitors Employing Junction Tree Variational Autoencoder with Bayesian Optimization and Gradient Ascent. *ACS Omega* **2024**, *9*, 47180–47193. [CrossRef]
37. Kondratyev, V.; Dryzhakov, M.; Gimadiev, T.; Slutskiy, D. Generative model based on junction tree variational autoencoder for HOMO value prediction and molecular optimization. *J. Cheminform.* **2023**, *15*, 11. [CrossRef]
38. Berman, H.M.; Westbrook, J.; Feng, Z.; Gilliland, G.; Bhat, T.N.; Weissig, H.; Shindyalov, I.N.; Bourne, P.E. The Protein Data Bank. *Nucleic Acids Res.* **2000**, *28*, 235–242. [CrossRef]
39. Lee, M.; Yau, W.-M.; Louis, J.M.; Tycko, R. Structures of brain-derived 42-residue amyloid-β fibril polymorphs with unusual molecular conformations and intermolecular interactions. *Proc. Natl. Acad. Sci. USA* **2023**, *120*, e2218831120. [CrossRef]
40. Dey, A.; Verma, A.; Bhaskar, U.; Sarkar, B.; Kallianpur, M.; Vishvakarma, V.; Das, A.K.; Garai, K.; Mukherjee, O.; Ishii, K.; et al. A Toxicogenic Interaction between Intracellular Amyloid-beta and Apolipoprotein-E. *ACS Chem. Neurosci.* **2024**, *15*, 1265–1275. [CrossRef]
41. Bhachoo, J.; Beuming, T. Investigating Protein-Peptide Interactions Using the Schrodinger Computational Suite. *Methods Mol. Biol.* **2017**, *1561*, 235–254. [CrossRef]
42. Hsu, P.W.; Lin, L.Z.; Hsu, S.D.; Hsu, J.B.; Huang, H.D. ViTa: Prediction of host microRNAs targets on viruses. *Nucleic Acids Res.* **2007**, *35*, D381–D385. [CrossRef]
43. Zdrazil, B.; Felix, E.; Hunter, F.; Manners, E.J.; Blackshaw, J.; Corbett, S.; de Veij, M.; Ioannidis, H.; Lopez, D.M.; Mosquera, J.F.; et al. The ChEMBL Database in 2023: A drug discovery platform spanning multiple bioactivity data types and time periods. *Nucleic Acids Res.* **2024**, *52*, D1180–D1192. [CrossRef]
44. Shaikh, F.; Siu, S.W. Identification of novel natural compound inhibitors for human complement component 5a receptor by homology modeling and virtual screening. *Med. Chem. Res.* **2016**, *25*, 1564–1573. [CrossRef]
45. Jin, W.; Barzilay, R.; Jaakkola, T.S. Junction Tree Variational Autoencoder for Molecular Graph Generation. *CoRR* **2018**, abs/1802.04364.
46. Jin, W.; Barzilay, R.; Jaakkola, T. Junction Tree Variational Autoencoder for Molecular Graph Generation. In Proceedings of the 35th International Conference on Machine Learning, Stockholm, Sweden, 10–15 July 2018; pp. 2323–2332.
47. Karauda, T.; Kornicki, K.; Jarri, A.; Antczak, A.; Milkowska-Dymanowska, J.; Piotrowski, W.J.; Majewski, S.; Gorski, P.; Bialas, A.J. Eosinopenia and neutrophil-to-lymphocyte count ratio as prognostic factors in exacerbation of COPD. *Sci. Rep.* **2021**, *11*, 4804. [CrossRef]
48. Thireou, T.; Reczko, M. Bidirectional Long Short-Term Memory Networks for predicting the subcellular localization of eukaryotic proteins. *IEEE/ACM Trans. Comput. Biol. Bioinform.* **2007**, *4*, 441–446. [CrossRef] [PubMed]
49. Kingma, D.P.; Welling, M. Auto-Encoding Variational Bayes. *arXiv* **2022**, arXiv:1312.6114.
50. Csiszar, I. I-Divergence Geometry of Probability Distributions and Minimization Problems. *Ann. Probab.* **1975**, *3*, 146–158. [CrossRef]
51. Colaboratory, G. Jupyter Notebook. Available online: https://colab.research.google.com (accessed on 10 November 2020).

biomolecules

MDPI

Article

GramSeq-DTA: A Grammar-Based Drug–Target Affinity Prediction Approach Fusing Gene Expression Information

Kusal Debnath [1], Pratip Rana [2] and Preetam Ghosh [1,*]

[1] Department of Computer Science, Virginia Commonwealth University, Richmond, VA 23284, USA; debnathk@vcu.edu
[2] Department of Computer Science, Old Dominion University, Norfolk, VA 23529, USA; prana@odu.edu
* Correspondence: pghosh@vcu.edu

Abstract: Drug–target affinity (DTA) prediction is a critical aspect of drug discovery. The meaningful representation of drugs and targets is crucial for accurate prediction. Using 1D string-based representations for drugs and targets is a common approach that has demonstrated good results in drug–target affinity prediction. However, these approach lacks information on the relative position of the atoms and bonds. To address this limitation, graph-based representations have been used to some extent. However, solely considering the structural aspect of drugs and targets may be insufficient for accurate DTA prediction. Integrating the functional aspect of these drugs at the genetic level can enhance the prediction capability of the models. To fill this gap, we propose GramSeq-DTA, which integrates chemical perturbation information with the structural information of drugs and targets. We applied a Grammar Variational Autoencoder (GVAE) for drug feature extraction and utilized two different approaches for protein feature extraction as follows: a Convolutional Neural Network (CNN) and a Recurrent Neural Network (RNN). The chemical perturbation data are obtained from the L1000 project, which provides information on the up-regulation and down-regulation of genes caused by selected drugs. This chemical perturbation information is processed, and a compact dataset is prepared, serving as the functional feature set of the drugs. By integrating the drug, gene, and target features in the model, our approach outperforms the current state-of-the-art DTA prediction models when validated on widely used DTA datasets (BindingDB, Davis, and KIBA). This work provides a novel and practical approach to DTA prediction by merging the structural and functional aspects of biological entities, and it encourages further research in multi-modal DTA prediction.

Keywords: drug–target affinity; deep learning; grammar-based encoding; chemical perturbation; multi-modal

1. Introduction

Drug–target affinity (DTA) prediction provides a foundation for modern drug discovery, with various benefits that improve efficiency, reduce costs, and increase success rates. Affinity reflects the likelihood of interaction of a drug–target pair, typically measured by K_d (dissociation constant) or K_i (inhibition constant). Strong binding is indicated by low K_d/K_i (nM to pM), while weak binding corresponds to high K_d/K_i (μM to mM). The significance of DTA prediction is well-discussed in the literature, emphasizing its role in accelerating the identification of potential drug candidates and minimizing the risk of failure during clinical trials [1,2]. Recent improvements in computational methods [3–6] and the availability of relevant data enhance the accuracy and reliability of DTA predictions [7,8], aiding in the design of efficient therapeutic strategies and effective treatments for many diseases.

To achieve accurate DTA predictions, the way in which both drugs and targets are represented is a critical determinant of the performance of the models. The proper encoding of these molecular entities is essential for capturing the intricate relationships between their structural and functional properties. Early computational approaches to DTA often relied on simplified representations, such as molecular fingerprints for drugs and amino acid sequences for proteins, to feed machine learning models. Although these methods showed some promise, their inability to fully capture the complexity of molecular interactions limited the predictive power of DTA models [9–11]. As the field evolved, researchers began exploring more sophisticated techniques to better model the structural and chemical properties of drugs and targets, leading to the development of advanced representation methods that significantly improved the predictive accuracy and generalizability of DTA models [12–14].

Ozturk et al. [15] proposed DeepDTA, where they utilized 1D convolutional neural networks (CNNs) to extract high-level representations of protein sequences and 1D SMILES representations of the compounds. Before this approach, most computational methods treated drug–target affinity prediction as a binary classification problem. DeepDTA redefined the problem as a continuum of binding strength values, providing a broader view of drug–target interactions.

Nguyen et al. [16] advanced the field by representing drugs as graphs instead of linear strings and utilized graph neural networks (GNNs) to predict drug–target affinity in their proposed deep learning model called GraphDTA. Building on this trend, Tran et al. [17] proposed the Deep Neural Computation (DeepNC) model, which consists of multiple graph neural network algorithms.

The rise of natural language-based methods in biomedical research has led to further innovations in DTA modeling [18–21]. Qiu et al. [22] introduced G-K-BertDTA to bridge the gap between the structural and semantic information of molecules. In their approach, drugs were represented as graphs to learn their topological features, and a knowledge-based BERT model was incorporated to obtain the semantic embeddings of the structures, thereby enhancing feature information.

Nevertheless, there are some limitations to the above-mentioned approaches. Firstly, the information on the relative positions of the constituent atoms and bonds is often missing in the drug encoding approaches adopted in these models. In addition, the functional aspects of those drugs, which can provide relevant insights into their interaction with targets, were also not incorporated.

To address these limitations, we utilized the encoding approach for drugs known as the grammar variational autoencoder (GVAE) proposed by Kusner et al. [23]. The GVAE discusses the parse tree-based encoding of the drug entities, which allows learning from semantic properties and syntactic rules. This approach can learn a more consistent latent space representation in which entities with nearby representations decode to similar outputs. In addition, to incorporate the functional aspect of those drugs, we integrated the chemical perturbation information from the L1000 project [24]. In the L1000 project, various chemical entities have been used as perturbagens and were tested against multiple human cell lines, primarily linked to several types of cancers, to analyze their gene expression profile. We have utilized these gene expression signatures as the functional feature set for the drugs. Thus, the approach followed in this work utilizes the structural and functional representation of drugs, which enhances the drug–target affinity prediction and outperforms the current state-of-the-art methods.

The paper provides a thorough overview of the background research in Section 2, along with a detailed explanation of the methodology, dataset preparation, network architecture, and evaluation metrics in Section 3. Section 4 presents the results of the proposed method on commonly used benchmark datasets and compares its performance to the current state-of-the-art DTA prediction models. Finally, Section 5 addresses the limitations of the study and explores opportunities for future research advancements.

2. Background

2.1. Grammar Variational Autencoder

Gómez-Bombarelli et al. [25] used Gated Recurrent Units (GRUs) and Deep Convolutional Neural Networks (DCNNs) to develop a generative model for molecular entities based on SMILES strings. This model has the potential to encode and decode molecular entities through a continuous latent space, which aids in the exploration of novel molecules with desirable properties in this space. Nevertheless, one major drawback in using string-based representation for molecular entities is their fragility, i.e., minute alterations in the string can lead to a complete deviation from the original molecule, even corresponding to the generation of entirely invalid entities. James et al. [26] first proposed the concept of constructing grammar for chemical structures. According to this work, every valid discrete entity can be represented as a parse tree from the given grammar. The advantage of generating parse trees compared to texts is that it ensures the complete validity of the generated entities based on grammar. Thus, Kusner et al. [23] proposed the grammar variational autoencoder (GVAE), which encodes and decodes directly from these parse trees. This approach allows the GVAE to learn from syntactic rules as well as to learn semantic properties. Along with its ability to generate valid outputs, this approach can also learn a more coherent latent space representation in which entities with nearby representations decode to similar outputs.

2.1.1. Context-Free Grammar

Context-free grammar (CFG) is conventionally defined as a 4-tuple $G = (V, \Sigma, R, S)$, where V represents a finite set of non-terminal symbols; Σ represents a finite set of terminal symbols, disjoint from V; R represents a finite set of production rules; S is a unique non-terminal referred to as the start symbol; and G represents the grammar that describes a set of trees that can be formed by applying rules in R to leaf nodes until all leaf nodes become terminal symbols in Σ.

The rules R are technically defined as $\alpha \rightarrow \beta$ for $\alpha \in V$ and $\beta \in (V \cup \Sigma*)$, * denoting the Kleene closure. Practically, these rules are portrayed as a collection of mappings from a solitary non-terminal on the left-hand side in V to a sequence of symbols that can be either terminal or non-terminal by definition. These mappings can be seen as a rule for rewriting.

When a production rule is applied to a non-terminal symbol, it creates a tree structure where the symbols on the right-hand side of the production rule become child nodes for the left-hand side parent. These trees extend from each non-terminal symbol in V. The language of G is the set of all sequences of terminal symbols that can be generated by traversing the leaf nodes of a tree from left to right. A parse tree is a tree with its root at S and a sequence of terminal symbols as its leaf nodes. The prevalence of context-free languages in computer science is attributed, in part, to the existence of practical parsing algorithms.

2.1.2. Syntactic vs. Semantic Validity

A crucial aspect of grammar-based encoding is that the encoded molecules are syntactically valid, but the semantic validity of these molecules is a matter of discussion. There are several reasons behind this phenomenon as follows: (a) Some molecules produced by the grammar may be unstable or chemically invalid; for example, a carbon atom cannot make bonds with more than four atoms in a molecule as it has a valency of 4. Nevertheless, the grammar can produce this kind of molecule. (b) The assignment of ring-bond digits in SMILES is a non-context-free process. It needs to keep track of the order in which rings are encountered, along with the connectivity between the rings, which can only be determined from the local context of the string. For example, in naphthalene (c1ccc2c(c1)cccc2), the outer ring uses the digit '1', and the inner ring uses '2'. The digits are not nested but rather follow a specific order. (c) The GVAE can still produce an undetermined sequence if there are existing non-terminal symbols on the stack after processing all logit vectors.

2.2. L1000 Assay

The L1000 project [24] is part of the Library of Integrated Network-Based Cellular Signatures (LINCS) program funded by the National Institutes of Health (NIH). This program aims to catalog and analyze cellular responses to various perturbations to understand how these perturbations modulate cellular functions. This project efficiently manages chemical perturbagen data using a structured method, involving data generation, processing, and analysis. This project uses various chemical compounds, including FDA-approved drugs, experimental drugs, natural compounds, and other bioactive molecules, as perturbagens. Perturbagen selection mainly involves the possible connection of these chemicals to numerous biological pathways and disease cross-talks. Various human cell lines, primarily associated with several types of cancers, are chosen to ensure diversity in biological responses. The L1000 data can thus be used to identify potential new uses for existing drugs or to discover new candidate drugs. Moreover, novel hypotheses can be made that correspond to the possible effects of the new compounds by performing comparative analyses of the gene expression signatures of known drugs. The data also aid in understanding the underlying molecular mechanism of the diseases by showcasing the effect of different compounds in the alteration of gene expression related to disease pathways.

3. Materials and Methods

3.1. Datasets

3.1.1. Benchmark Datasets

In this study, the following three datasets are used for benchmarking purposes: BindingDB [27], Davis [28], and KIBA [29]. The drug–target affinity dataset in the BindingDB database contains experimental binding affinities between small molecules and protein targets and pharmacological annotations of the entities (e.g., ID, Structure, etc.). In the Davis dataset, the targets are kinase proteins, and the drugs are the small molecules (inhibitors) targeting those kinases. Similar to Davis, the KIBA dataset also focuses on kinase proteins and their corresponding inhibitor drugs, but it contains a more significant number of instances compared to Davis. The variations in drug and target components across these datasets, thus, allow for a comprehensive evaluation of the robustness and generalizability of the model.

In these datasets, the drugs are represented as SMILES strings, and the target proteins are represented as amino acid sequences. For the BindingDB and Davis datasets, the labels are the K_d (dissociation constant) values, which indicate the extent of the interactions between each drug–target pair. Meanwhile, a unified KIBA score is used as a label for the KIBA dataset, combining K_d, K_i (inhibition constant), and IC_{50} (half-maximal inhibitory coefficient) values for the corresponding drug–target pairs. While Davis and KIBA are readily available for training without any prior preprocessing, BindingDB contains the missing values among various features.

Thus, thorough preprocessing is performed on BindingDB, keeping only the SMILES target protein sequences and the K_d values as the required features, discarding everything else. Even if this preprocessing leads to a certain amount of data loss, this step eventually produces cleaner data suitable for model training. The labels are converted into logarithmic form which helps improve the performance of the model in regression tasks.

3.1.2. L1000 Chemical Perturbation Dataset

The chemical perturbation data available in the L1000 project is documented in raw format. Therefore, the data need to be processed accordingly for use. The detailed process of preparing the L1000 dataset from the raw data is discussed below.

①The L1000 chemical perturbation data file is loaded where each perturbagen has multiple replicates based on dosage concentration, and each replicate has two lists of associated genes—one for up-regulated and the other for down-regulated genes.②The analysis of the dosage concentration distribution among the replicates shows that samples with a concentration of 10 μM are the most common. Gene expression responses can be dose-dependent; thus, selecting single-dose samples provides more biological relevance, providing a standardized functional overview of the drug effects. Therefore, for standardization purposes, samples with a concentration of 10 μM are selected for further analysis, while the others are excluded.③ Each unique perturbagen is mapped into its corresponding SMILES representation, which is important for downstream molecular modeling.④For each perturbagen p_i, the gene regulatory information is represented as a vector of 'up'- and 'down'-regulation values across 978 landmark genes, and the number of times a gene is up-regulated or down-regulated is counted and normalized by the number of replicates. Let x_{ij}^{up} represent whether gene j is up-regulated for perturbagen p_i and x_{ij}^{down} represent the same for down-regulation. The final regulatory vector for each perturbagen is computed as follows:

$$\mathbf{v}_i = \frac{1}{\text{count}(p_i)} \left(\sum_{j=1}^{m} x_{ij}^{\text{up}}, \sum_{j=1}^{m} x_{ij}^{\text{down}} \right) \tag{1}$$

where $m = 978$ is the number of landmark genes, and $\text{count}(p_i)$ is the number of times perturbagen p_i appears in the dataset. A representative illustration is shown in Figure 1. ⑤ Finally, the vector for each perturbagen is stacked to obtain the final matrix as follows:

$$\mathbf{V} = \{\mathbf{v}_i\}_{i=1}^{k}, \quad \mathbf{v}_i \in \mathbb{R}^{978\times 2} \tag{2}$$

where $\mathbf{V}$ is of shape $k \times 978 \times 2$, and k is the number of unique perturbagens. The processed dataset is an important contribution of this work that can aid future research in RNA-Seq data integration and analysis.

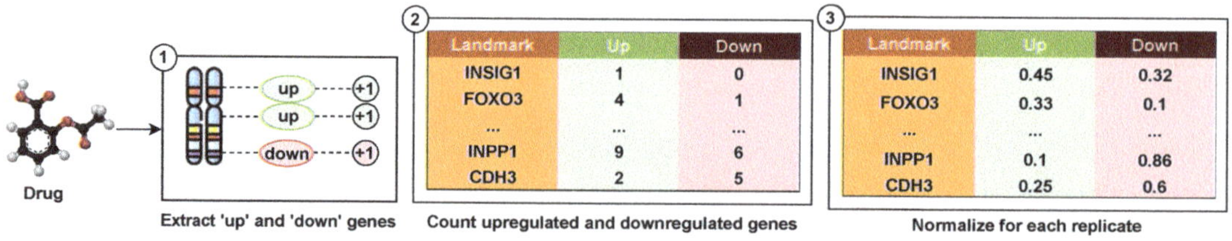

Figure 1. Preparation of the gene expression dataset. Gene expression information analyzed on 978 landmark genes for the selected drugs is extracted from the L1000 chemical perturbation data. After considering all the biological replicates of the perturbation analysis, a gene regulation matrix is created for both up-regulated and down-regulated genes.

The results from the L1000 assay highlight the potential to capture any cellular state by measuring the reduced representation of the human transcriptome. In that study, 12,031 Affymetrix HGU133A expression profiles from the Gene Expression Omnibus (GEO) [30] were analyzed. The analysis determined that the optimal number of informative transcripts (k) is 978. Selecting a lower value for k results in significant information loss, while higher values fail to offer a substantial cost reduction compared to the entire transcriptome. Ultimately, these 978 selected landmarks were sufficient to represent 82% of the complete transcriptomic information. Thus, for the construction of the chemical perturbation dataset here, that list of 978 genes was used.

Not all the perturbagens mentioned in the L1000 dataset are entirely present in the benchmark datasets. Therefore, the datasets are processed accordingly, and only those drugs whose corresponding regulatory vector is present in the L1000 dataset are selected. The processing of the datasets resulted in a decrease in the number of total interactions. A summary of all the original and processed benchmark datasets is presented in Table 1.

Table 1. Dataset statistics—number of compounds, proteins, and interactions.

Dataset	Compounds	Proteins	Interactions
Original			
BindingDB	22,381	1860	91,751
Davis	68	379	30,056
KIBA	2068	229	118,254
Processed			
BindingDB	444	754	18,567
Davis	11	379	4169
KIBA	12	194	538

3.2. Network Architecture

The complete network consists of four main parts: (a) a drug encoder, (b) RNA-Seq encoder, (c) protein encoder and (d) regression head. Figure 2 shows the schematic diagram of the complete network architecture.

3.2.1. Drug Encoder

For this work, we utilized a pre-trained GVAE model from the study conducted by Zhu et al. [31] that focuses on deep learning-based drug efficacy prediction from transcriptional profiles. One-hot encoded vectors are generated by parsing the SMILESs representations of the drugs using a grammar-tree-based approach and subsequently passing those into the encoder network. The detailed process of parsing the SMILES and generating one-hot encoded vectors is discussed below.

① RDKit [32] tool is used to canonize all SMILESs to a standardized representation of SMILESs, as the same molecule collected from different sources may possess different SMILES representations. ② Canonized SMILES representations are converted into a collection of tokens using a tokenizer. ③ The tokenized sequence is then parsed using a grammar adopted from the work of Kusner et al. [23]. This yields a sequence of production rules as follows:

$$G(\tau(S)) = P = \{P_1, P_2, \ldots, P_q\} \tag{3}$$

where G is the grammar, $\tau(S)$ is the tokenized sequence, and $P = \{P_1, P_2, \ldots, P_q\}$ is the sequence of production rules. Each production rule is then mapped to an index I_i in a predefined list of rules. ④ A zero matrix is initialized, denoting the vector to be populated by one-hot encoding as follows:

$$O_{j,I_j} = 1, \quad \forall j \in \{1,2,\ldots,\min(M,q)\}, \quad I_j \in \{1,2,\ldots,N-1\} \tag{4}$$

where O is the encoded one-hot vector of shape $M \times N$, M is the maximum length of sequences, N is the total number of production rules, and q is the number of productions. If q is smaller than M, the rest of the matrix is padded with an indicator for "end of sequence", as follows:

$$O_{j,N-1} = 1, \quad \forall j \in \{q+1, q+2, \ldots, M\} \tag{5}$$

In this work, the values of M and N are 277 and 76, respectively. The generated one-hot vectors for each SMILES representations are then passed into the encoder network. The schematic diagram of the overall process is given in Figure 3.

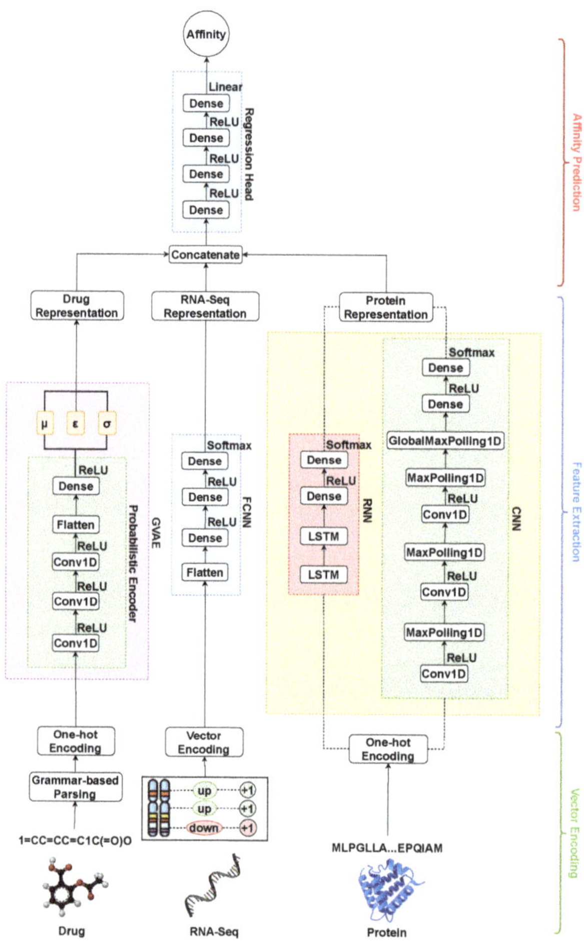

Figure 2. Network architecture of the proposed model. The encoded drug information is passed through an GVAE block, the RNA-Seq information is passed through an FCNN, while the encoded protein information is passed through a series of LSTM layers and 1D CNN layers. Learned representations are concatenated and passed through a FCNN acting as a regression head to predict the affinity.

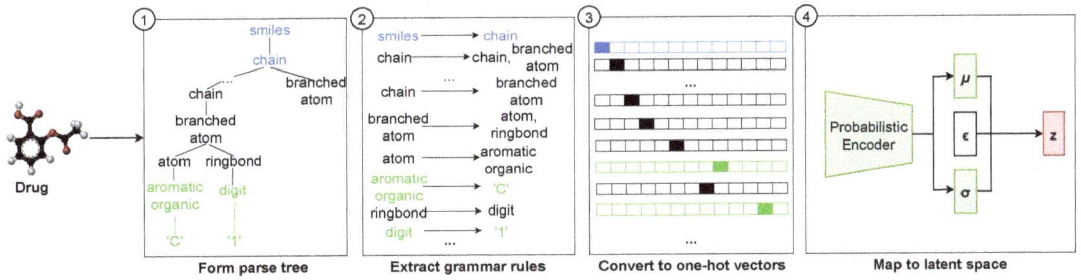

Figure 3. Encoding of drug SMILES structures. A parse tree is constructed based on the structural components of SMILES representations. Grammar rules are extracted from the parsed trees. SMILES representations are then converted into one-hot vectors. Finally, the one-hot vectors are transformed into the corresponding latent space representations using an encoder network.

3.2.2. RNA-Seq Encoder

A fully connected neural network (FCNN) [33] is used for the extraction of meaningful features from the high-dimensional RNA-Seq data. As discussed in the L1000 chemical perturbation dataset preparation, the resulting dataset is of $k \times 978 \times 2$ shape, where k is the number of unique perturbagens, 978 is the number of landmark genes, and 2 indicates the number of columns representing up-regulated and down-regulated genes. When these vectors are passed through a fully connected neural network, it learns a condensed and abstract representation of how each perturbagen affects gene expression.

3.2.3. Protein Encoder

Feature extraction from amino acid sequences is achieved using two different types of neural networks as follows: a Convolutional Neural Network (CNN) and a Recurrent Neural Network (RNN) [33]. CNNs are able to identify local motifs and patterns within a sequence by using the sliding windows of filters to capture neighboring amino acids. On the other hand, RNNs are capable of extracting long-range dependencies and sequential relationships between amino acids by retaining information from previous positions in the sequence in a step-by-step manner. To encode the sequences, a dictionary of all possible amino acid symbols in the proteins is created. One-hot encoding of a given protein is carried out based on the presence of a particular amino acid in that protein. For standardization, the maximum length of a protein is limited to 1000 sequences. The encoding of all the proteins results in the creation of a tensor of $p \times 26 \times 1000$, where p is the number of unique proteins and 26 is the length of the amino acid dictionary.

3.3. Training Settings

A pre-trained GVAE model from Zhu et al. [31] is used as the drug encoder for deep learning-based drug efficacy prediction from transcriptional profiles. Since the model is already pre-trained, no additional training is needed for this encoder block. It generates a latent representation for the drugs, which is then concatenated with representations extracted separately—by the FCNN block for RNA-Seq data and by the CNN and RNN blocks for protein data. Finally, these combined representations are fed into a regression head, consisting of fully connected layers, to predict affinity. Unlike the drug encoder, the RNA-Seq and protein encoders, as well as the regression head, are trained from scratch. The combined network is trained together in an end-to-end manner. The training process is set to run for 500 epochs with an adaptive learning rate that starts at 0.001 using the Adam optimizer. The higher learning rate value is chosen to ensure that the training process does not significantly impact the pre-trained weights in the GVAE model. A batch size of 256 has been found to yield the best results, maintaining a balance between memory usage and convergence speed. A summary of the overall network architecture and training hyperparameters are presented in Table 2.

Table 2. Summary of network architectures and training hyperparameters.

Parameters	Value
Drug Encoding	
GVAE Encoder Filter Sizes	9, 9, 10
GVAE Encoder Kernel Sizes	9, 9, 11
GVAE Latent Space Dimension	56
RNA-Seq Encoding	
Dense Layers	2
Protein Encoding	
CNN Filter Sizes	32, 64, 96
CNN Kernel Sizes	4, 8, 12
RNN LSTM Layers	2
Regression Head	
Dense Layers	3
Training	
Epochs	500
Learning Rate	0.001
Batch Size	256
Optimizer	Adam

3.4. Evaluation Metrics

Evaluating deep learning models involves various metrics that capture different aspects of the performance. Mean Squared Error (MSE) measures the average squared difference between the actual and the predicted values, highlighting prediction accuracy. We also used the Concordance Index (CI), which is a preferred metric for the survival analysis, to evaluate the consistency between the predicted risk scores and the actual outcomes. Together, these metrics provide a robust framework for model evaluation.

3.4.1. Mean Squared Error (MSE)

The Mean Squared Error (MSE) is a common loss function used for regression tasks. It measures the average of the squares of the errors, which are the differences between the predicted and actual values. Mathematically, it is defined as follows:

$$\text{MSE} = \frac{1}{n}\sum_{i=1}^{n}(y_i - \hat{y}_i)^2 \tag{6}$$

For datasets BindingDB and Davis, y_i is the actual negative log of the K_d value, $\hat{y}_i$ is the predicted negative log of the K_d value, while K_d is replaced by the KIBA score for the KIBA dataset. n is the number of data points.

3.4.2. Concordance Index (CI)

The Concordance Index (CI) is a metric used primarily in survival analysis to evaluate the predictive accuracy of risk scores. It assesses the degree of concordance between the predicted and actual ordering of event times. The CI is able to measure how often a randomly chosen pair is ranked correctly with respect to the ground truth, which is particularly relevant for ranking-based problems like drug–target affinity prediction. The CI is calculated as follows:

$$CI = \frac{\text{Number of concordant pairs}}{\text{Number of possible evaluation pairs}} \quad (7)$$

A pair is considered concordant if the predicted and actual orderings of two instances are consistent.

4. Results and Discussion

In this section, we will discuss the performance of the GramSeq-DTA model in detail. The discussion can be divided into the following parts: (a) benchmarking the performance of GramSeq-DTA with integrated RNA-Seq information against the baseline models, (b) the advantage of integrating RNA-Seq information to the model, and (c) performance comparison of the proposed model using original and processed datasets.

4.1. Benchmarking Against Baseline Models

In order to validate our findings, we conducted a comprehensive performance comparison of GramSeq-DTA, which now includes integrated RNA-Seq information. We compared it against several well-established baseline models as follows: DeepDTA, GraphDTA, DeepNC, and G-K-BertDTA. We used the MSE and CI values to assess performance. The model performance was compared across three benchmark datasets—BindingDB, Davis, and KIBA. DeepDTA employs a deep learning framework to capture the complex features of drug–target interactions using convolutional neural networks. GraphDTA, on the other hand, leverages graph neural networks to represent molecular structures as graphs, enabling it to better capture the topological properties of molecules. DeepNC uses a neural collaborative filtering approach to model interactions, focusing on latent feature extraction. Lastly, G-K-BertDTA integrates graph-based representations with BERT-like architectures to enhance the contextual understanding of molecular relationships. Our extensive evaluation, conducted on processed benchmark datasets, showed that GramSeq-DTA consistently outperformed its counterparts in terms of Concordance Index (CI) values, a widely accepted metric for evaluating predictive performance in drug–target affinity modeling. In the BindingDB dataset, GramSeq-DTA outperforms G-K-BertDTA by 1.32% in terms of CI value. Similarly, on the Davis dataset, GramSeq-DTA shows a 0.89% advantage over DeepNC. On the KIBA dataset, GramSeq-DTA exceeds G-K-BertDTA by 2.75%. Importantly, the integration of RNA-Seq data into GramSeq-DTA provided valuable insights into gene expression patterns, contributing to the improved accuracy of the models in predicting drug–target interactions. Detailed results of this comparison can be found in Tables 3–5, where the enhanced GramSeq-DTA model demonstrates its robust performance, setting a new standard in the field.

Table 3. Performance comparison of GramSeq-DTA with baseline models on the processed BindingDB dataset.

Model	Drug Encoder	RNA-Seq Encoder	Protein Encoder	MSE	CI
DeepDTA	CNN	-	CNN	0.384	0.821
GraphDTA	GINConvNet	-	CNN	0.355	0.819
	GCNNet	-	CNN	0.397	0.786
	GATNet	-	CNN	0.512	0.757
	GAT_GCN	-	CNN	0.384	0.806
DeepNC	GENConv	-	CNN	0.367	0.828
G-K-BertDTA	GINConvNet + Embeddings	-	CNN	0.325	0.832
GramSeq-DTA	GVAE	FCNN	CNN	0.365	0.843
	GVAE	FCNN	RNN	0.355	0.832

Table 4. Performance comparison of GramSeq-DTA with baseline models on the processed Davis dataset.

Model	Drug Encoder	RNA-Seq Encoder	Protein Encoder	MSE	CI
DeepDTA	CNN	-	CNN	0.219	0.779
GraphDTA	GINConvNet	-	CNN	0.187	0.771
	GCNNet	-	CNN	0.214	0.732
	GATNet	-	CNN	0.241	0.713
	GAT_GCN	-	CNN	0.227	0.753
DeepNC	GENConv	-	CNN	0.198	0.789
G-K-BertDTA	GINConvNet + Embeddings	-	CNN	0.169	0.778
GramSeq-DTA	GVAE	FCNN	CNN	0.293	0.796
	GVAE	FCNN	RNN	0.261	0.796

Table 5. Performance comparison of GramSeq-DTA with baseline models on the processed KIBA dataset.

Model	Drug Encoder	RNA-Seq Encoder	Protein Encoder	MSE	CI
DeepDTA	CNN	-	CNN	0.877	0.609
GraphDTA	GINConvNet	-	CNN	1.061	0.628
	GCNNet	-	CNN	0.903	0.631
	GATNet	-	CNN	0.957	0.609
	GAT_GCN	-	CNN	0.831	0.671
DeepNC	GENConv	-	CNN	0.769	0.648
G-K-BertDTA	GINConvNet + Embeddings	-	CNN	0.693	0.689
GramSeq-DTA	GVAE	FCNN	CNN	0.843	0.708
	GVAE	FCNN	RNN	1.269	0.688

4.2. Advantage of Integrating RNA-Seq Information

Table 6 indicates that when validated on the processed BindingDB dataset, GramSeq-DTA performs better, with a CI value of 0.843 when integrating RNA-Seq information compared to not integrating RNA-Seq information. Similar results are evident in Tables 7 and 8, where validation is performed on the Davis and KIBA datasets, respectively. CI values of 0.796 and 0.708 are observed for the processed Davis and KIBA datasets, respectively, when RNA-Seq information is integrated. These observations prove that integrating RNA-Seq information with the corresponding drug and target structural information can enhance the drug–target affinity prediction ability of the model.

Table 6. Performance comparison of GramSeq-DTA with and without RNA-Seq information integration on the processed BindingDB dataset.

Model	Drug Encoder	RNA-Seq Encoder	Protein Encoder	MSE	CI
GramSeq-DTA	GVAE	-	CNN	0.495	0.746
	GVAE	-	RNN	0.495	0.754
	GVAE	FCNN	CNN	0.365	0.843
	GVAE	FCNN	RNN	0.355	0.832

Table 7. Performance comparison of GramSeq-DTA with and without RNA-Seq information integration on the processed Davis dataset.

Model	Drug Encoder	RNA-Seq Encoder	Protein Encoder	MSE	CI
GramSeq-DTA	GVAE	-	CNN	0.277	0.705
	GVAE	-	RNN	0.311	0.716
	GVAE	FCNN	CNN	0.293	0.796
	GVAE	FCNN	RNN	0.261	0.796

Table 8. Performance comparison of GramSeq-DTA with and without RNA-Seq information integration on the processed KIBA dataset.

Model	Drug Encoder	RNA-Seq Encoder	Protein Encoder	MSE	CI
GramSeq-DTA	GVAE	-	CNN	1.011	0.653
	GVAE	-	RNN	0.876	0.618
	GVAE	FCNN	CNN	0.843	0.708
	GVAE	FCNN	RNN	1.269	0.688

4.3. Performance on Original and Processed Data

Table 9 presents the comparative evaluation of the performance of the proposed model on the original benchmark datasets and the processed benchmark datasets. The results of the model integration with RNA-Seq information are shown for the processed datasets. As shown in Table 1, the number of interactions between the original and the processed datasets differs. Despite losing approximately 80% of data in processing, our model performs better on the processed BindingDB dataset, with a best CI value of 0.843, while for the original BindingDB dataset, the optimum CI value was 0.818. The results on the original and processed Davis dataset are also competitive. When processing the Davis dataset, we lost around 84% of data. Our model indicates a CI value of 0.809 for the original Davis dataset, while for the processed Davis dataset, the CI value is 0.796. Data loss during the processing of the KIBA dataset is 99.5%, which is the highest value among all three datasets. For the KIBA dataset, there is a significant difference in CI values (original: 0.823; processed: 0.708) for original and processed datasets. The performance difference for the KIBA dataset can be due to the excessive data loss while processing the original dataset. This data loss leads to reduced data diversity and thus affects the overall generalizability of the model while training. Based on the performance of the other two process datasets (BindingDB and Davis), with more data available for the processed KIBA dataset, there is a high possibility of achieving better performance.

Table 9. Performance of GramSeq-DTA without RNA-Seq information integration on the original benchmark datasets.

Dataset	Model	Drug Encoder	RNA-Seq Encoder	Protein Encoder	MSE	CI
BindingDB						
Original	GramSeq-DTA	GVAE	-	CNN	1.029	0.818
		GVAE	-	RNN	1.061	0.812
Processed	GramSeq-DTA	GVAE	FCNN	CNN	0.365	0.843
		GVAE	FCNN	RNN	0.355	0.832
Davis						
Original	GramSeq-DTA	GVAE	-	CNN	0.446	0.806
		GVAE	-	RNN	0.445	0.809
Processed	GramSeq-DTA	GVAE	FCNN	CNN	0.293	0.796
		GVAE	FCNN	RNN	0.261	0.796
KIBA						
Original	GramSeq-DTA	GVAE	-	CNN	0.272	0.823
		GVAE	-	RNN	0.277	0.823
Processed	GramSeq-DTA	GVAE	FCNN	CNN	0.843	0.708
		GVAE	FCNN	RNN	1.269	0.688

5. Conclusions and Future Directions

In this study, we have demonstrated that incorporating chemical perturbation information can enhance drug–target affinity prediction. The core contribution of this research is in its transformation of data from a chemical perturbation assay and utilizing them as an extra modality along with drug and protein structural information. In this work, we transformed the raw chemical perturbation assay data from the L1000 project, a project where a vast array of chemical assays have been performed, and updates to the assays are still ongoing. This research could guide future work on better understanding affinity prediction among biological entities in the absence of the three-dimensional structural information of those entities.

Nevertheless, a fundamental limitation of this research is that information from a chemical perturbation assay may not be available for every drug in the widely used drug–target affinity benchmark datasets. As new drugs are being introduced as potent antidotes for various diseases, corresponding chemical perturbation studies need to be performed on a large scale. In addition, some drugs are also being discarded and prohibited for conventional usage due to the lack of efficacy and severe side effects. Another aspect to consider is the imbalance between active and inactive compounds in a dataset. Chemical perturbation data might help by adding meaningful biological signals, but they could also worsen things if they introduce too much noise. Processed datasets can reduce the imbalance by filtering out redundant inactive data. However, if not handled carefully, they might remove important active compounds or introduce biases that do not reflect real-world data. Future studies should investigate different approaches for transforming chemical perturbation information.

Recent studies have shown that representing drugs as graphs provides a better understanding of the positional aspects of interacting atoms within the compounds, providing deeper insights into the structure–activity relationship (SAR) of the compounds. In addition, the recent availability of predicted 3D structures for the target proteins provides an opportunity to integrate these 3D structural data into the model, therefore enhancing the prediction capabilities of the computational models. This integration can provide deeper insights into the relationship between the structure and function of the proteins, providing a better understanding of structure-based predictive DTA modeling. Moreover, introducing advanced feature extraction methods from biological entities can enhance prediction.

With the continuous developments in natural language-based approaches in deep learning, drug discovery and drug repurposing are fields where the application of such methodologies are becoming popular. As the drugs and proteins are represented as strings, the utilization of natural language-based approaches can enhance the prediction abilities of the computational models. In the future, a conjugation of the existing structure-based approaches and natural language-based approaches can aid in improving the model development for DTA and the related tasks in drug discovery.

In summary, this work underscores the importance of integrating additional data modalities in drug–target affinity prediction.

Author Contributions: Conceptualization, K.D. and P.R.; methodology, K.D.; formal analysis, K.D.; investigation, P.R.; resources, P.G.; data curation, K.D.; writing—original draft preparation, K.D.; writing—review and editing, K.D., P.R. and P.G.; visualization, K.D.; supervision, P.G.; project administration, P.G.; funding acquisition, P.G. All authors have read and agreed to the published version of the manuscript.

Funding: This work was partially supported by the 5R21MH128562-02 (PI: Roberson-Nay), 5R21AA029492-02 (PI: Roberson-Nay), CHRB-2360623 (PI: Das), NSF-2316003 (PI: Cano), VCU Quest (PI: Das), and VCU Breakthroughs (PI: Ghosh) funds awarded to P.G.

Institutional Review Board Statement: Not applicable.

Informed Consent Statement: Not applicable.

Data Availability Statement: The data that support the findings of our study can be found in the GitHub repository, https://github.com/debnathk/gramseq.git (accessed on 16 January 2025).

Conflicts of Interest: The authors declare no conflict of interest.

References

1. Mak, K.K.; Wong, Y.H.; Pichika, M.R. Artificial intelligence in drug discovery and development. *Drug Discov. Today* **2023**, *26*, 80–93. [CrossRef]
2. Kim, J.; Park, S.; Min, D.; Kim, W. Comprehensive survey of recent drug discovery using deep learning. *Int. J. Mol. Sci.* **2021**, *22*, 9983. [CrossRef] [PubMed]
3. Chaitankar, V.; Ghosh, P.; Perkins, E.; Gong, P.; Deng, Y.; Zhang, C. A novel gene network inference algorithm using predictive minimum description length approach. *BMC Syst. Biol.* **2010**, *4*, S7. [CrossRef]
4. Chaitankar, V.; Ghosh, P.; Perkins, E.; Gong, P.; Zhang, C. Time lagged information-theoretic approaches to the reverse engineering of gene regulatory networks. *BMC Bioinform.* **2010**, *11*, S19. [CrossRef]
5. Nalluri, J.J.; Barh, D.; Azevedo, V.; Ghosh, P. Mirsig: A consensus-based network inference methodology to identify pan-cancer Mirna-MIRNA interaction signatures. *Sci. Rep.* **2017**, *7*, 39684. [CrossRef]
6. Franco, E.F.; Rana, P.; Cruz, A.; Calderón, V.V.; Azevedo, V.; Ramos, R.T.J.; Ghosh, P. Performance Comparison of Deep Learning Autoencoders for Cancer Subtype Detection Using Multi-Omics Data. *Cancers* **2021**, *13*, 2013. [CrossRef] [PubMed]
7. Meli, R.; Morris, G.M.; Biggin, P.C. Scoring functions for protein-ligand binding affinity prediction using structure-based deep learning: A review. *Front. Bioinform.* **2022**, *2*, 885983. [CrossRef]
8. Moon, S.; Zhung, W.; Kim, W.Y. Toward generalizable structure-based deep learning models for protein–ligand interaction prediction: Challenges and strategies. *Wiley Interdiscip. Rev. Comput. Mol. Sci.* **2024**, *14*, e1705. [CrossRef]
9. Ding, H.; Takigawa, I.; Mamitsuka, H.; Zhu, S. Similarity-based machine learning methods for predicting drug–target interactions: A brief review. *Briefings Bioinform.* **2014**, *15*, 734–747. [CrossRef]
10. Pahikkala, T.; Airola, A.; Pietilä, S.; Shakyawar, S.; Szwajda, A.; Tang, J.; Aittokallio, T. Toward more realistic drug–target interaction predictions. *Briefings Bioinform.* **2015**, *16*, 325–337. [CrossRef]
11. He, T.; Heidemeyer, M.; Ban, F.; Cherkasov, A.; Ester, M. SimBoost: A read-across approach for predicting drug–target binding affinities using gradient boosting machines. *J. Cheminform.* **2017**, *9*, 24. [CrossRef] [PubMed]
12. Ru, X.; Ye, X.; Sakurai, T.; Zou, Q.; Xu, L.; Lin, C. Current status and future prospects of drug—Target interaction prediction. *Briefings Funct. Genom.* **2021**, *20*, 312–322. [CrossRef]
13. Chen, M.; Jiang, Y.; Lei, X.; Pan, Y.; Ji, C.; Jiang, W.; Xiong, H. Drug-target interactions prediction based on signed heterogeneous graph neural networks. *Chin. J. Electron.* **2024**, *33*, 231–244. [CrossRef]
14. Abdul Raheem, A.K.; Dhannoon, B.N. Comprehensive Review on Drug-target Interaction Prediction-Latest Developments and Overview. *Curr. Drug Discov. Technol.* **2024**, *21*, 56–67. [CrossRef] [PubMed]
15. Öztürk, H.; Özgür, A.; Ozkirimli, E. DeepDTA: Deep drug–target binding affinity prediction. *Bioinformatics* **2018**, *34*, i821–i829. [CrossRef]
16. Nguyen, T.; Le, H.; Quinn, T.P.; Nguyen, T.; Le, T.D.; Venkatesh, S. GraphDTA: Predicting drug–target binding affinity with graph neural networks. *Bioinformatics* **2021**, *37*, 1140–1147. [CrossRef]
17. Tran, H.N.T.; Thomas, J.J.; Malim, N.H.A.H. DeepNC: A framework for drug-target interaction prediction with graph neural networks. *PeerJ* **2022**, *10*, e13163. [CrossRef]
18. Wolf, T.; Debut, L.; Sanh, V.; Chaumond, J.; Delangue, C.; Moi, A.; Cistac, P.; Rault, T.; Louf, R.; Funtowicz, M.; et al. HuggingFace's Transformers: State-of-the-art Natural Language Processing. *arXiv* **2019**, arXiv:1910.03771.
19. Bhatnagar, R.; Sardar, S.; Beheshti, M.; Podichetty, J.T. How can natural language processing help model informed drug development?: A review. *JAMIA Open* **2022**, *5*, ooac043. [CrossRef]
20. Grisoni, F. Chemical language models for de novo drug design: Challenges and opportunities. *Curr. Opin. Struct. Biol.* **2023**, *79*, 102527. [CrossRef]

21. Le, N.Q.K. Leveraging transformers-based language models in proteome bioinformatics. *Proteomics* **2023**, *23*, 2300011. [CrossRef] [PubMed]
22. Qiu, X.; Wang, H.; Tan, X.; Fang, Z. GK BertDTA: A graph representation learning and semantic embedding-based framework for drug-target affinity prediction. *Comput. Biol. Med.* **2024**, *173*, 108376. [CrossRef]
23. Kusner, M.J.; Paige, B.; Hernández-Lobato, J.M. Grammar Variational Autoencoder. In Proceedings of the 34th International Conference on Machine Learning, Sydney, Australia, 6–11 August 2017; Volume 70, pp. 1945–1954.
24. Subramanian, A.; Narayan, R.; Corsello, S.M.; Peck, D.D.; Natoli, T.E.; Lu, X.; Gould, J.; Davis, J.F.; Tubelli, A.A.; Asiedu, J.K.; et al. A next generation connectivity map: L1000 platform and the first 1,000,000 profiles. *Cell* **2017**, *171*, 1437–1452. [CrossRef] [PubMed]
25. Gómez-Bombarelli, R.; Wei, J.N.; Duvenaud, D.; Hernández-Lobato, J.M.; Sánchez-Lengeling, B.; Sheberla, D.; Aguilera-Iparraguirre, J.; Hirzel, T.D.; Adams, R.P.; Aspuru-Guzik, A. Automatic chemical design using a data-driven continuous representation of molecules. *ACS Cent. Sci.* **2018**, *4*, 268–276. [CrossRef]
26. OpenSMILES. The OpenSMILES Format. 2015. Available online: http://opensmiles.org/opensmiles.html (accessed on 2 October 2024).
27. Liu, T.; Lin, Y.; Wen, X.; Jorissen, R.N.; Gilson, M.K. BindingDB: A web-accessible database of experimentally determined protein–ligand binding affinities. *Nucleic Acids Res.* **2007**, *35*, D198–D201. [CrossRef] [PubMed]
28. Davis, M.I.; Hunt, J.P.; Herrgard, S.; Ciceri, P.; Wodicka, L.M.; Pallares, G.; Hocker, M.; Treiber, D.K.; Zarrinkar, P.P. Comprehensive analysis of kinase inhibitor selectivity. *Nat. Biotechnol.* **2011**, *29*, 1046–1051. [CrossRef]
29. Tang, J.; Szwajda, A.; Shakyawar, S.; Xu, T.; Hintsanen, P.; Wennerberg, K.; Aittokallio, T. Making sense of large-scale kinase inhibitor bioactivity data sets: A comparative and integrative analysis. *J. Chem. Inf. Model.* **2014**, *54*, 735–743. [CrossRef]
30. Clough, E.; Barrett, T. The gene expression omnibus database. In *Statistical Genomics: Methods and Protocols*; Humana Press: New York, NY, USA, 2016; pp. 93–110.
31. Zhu, J.; Wang, J.; Wang, X.; Gao, M.; Guo, B.; Gao, M.; Liu, J.; Yu, Y.; Wang, L.; Kong, W.; et al. Prediction of drug efficacy from transcriptional profiles with deep learning. *Nat. Biotechnol.* **2021**, *39*, 1444–1452. [CrossRef]
32. Landrum, G. RDKit: Open-Source Cheminformatics. 2024. Available online: https://www.rdkit.org/ (accessed on 10 February 2025).
33. Sainath, T.N.; Vinyals, O.; Senior, A.; Sak, H. Convolutional, long short-term memory, fully connected deep neural networks. In Proceedings of the 2015 IEEE International Conference on Acoustics, Speech and Signal Processing (ICASSP), Brisbane, Australia, 19–24 April 2015; pp. 4580–4584.

biomolecules

MDPI

Article

Insights into the Effects of Ligand Binding on Leucyl-tRNA Synthetase Inhibitors for Tuberculosis: In Silico Analysis and Isothermal Titration Calorimetry Validation

Zia Ur Rehman *, Asim Najmi and Khalid Zoghebi

Department of Pharmaceutical Chemistry, Faculty of Pharmacy, Jazan University, P.O. Box 114, Jazan 45142, Saudi Arabia; anajmi@jazanu.edu.sa (A.N.); kzoghebi@jazanu.edu.sa (K.Z.)
* Correspondence: zrehman@jazanu.edu.sa; Tel.: +966-550594249

Abstract: Incidences of drug-resistant tuberculosis have become common and are rising at an alarming rate. Aminoacyl t-RNA synthetase has been validated as a newer target against *Mycobacterium tuberculosis*. Leucyl t-RNA synthetase (LeuRS) is ubiquitously found in all organisms and regulates transcription, protein synthesis, mitochondrial RNA cleavage, and proofreading of matured t-RNA. Leucyl t-RNA synthetase promotes growth and development and is the key enzyme needed for biofilm formation in Mycobacterium. Inhibition of this enzyme could restrict the growth and development of the mycobacterial population. A database consisting of 2734 drug-like molecules was screened against leucyl t-RNA synthetase enzymes through virtual screening. Based on the docking scores and MMGBSA energy values, the top three compounds were selected for molecular dynamics simulation. The druggable nature of the top three hits was confirmed by predicting their pharmacokinetic parameters. The top three hits—compounds **1035** (ZINC000001543916), **1054** (ZINC000001554197), and **2077** (ZINC000008214483)—were evaluated for their binding affinity toward leucyl t-RNA synthetase by an isothermal titration calorimetry study. The inhibitory activity of these compounds was tested against antimycobacterial activity, biofilm formation, and LeuRS gene expression potential. Compound **1054** (Macimorelin) was found to be the most potent molecule, with better antimycobacterial activity, enzyme binding affinity, and significant inhibition of biofilm formation, as well as inhibition of the LeuRS gene expression. Compound **1054**, the top hit compound, has the potential to be used as a lead to develop successful leucyl t-RNA synthetase inhibitors.

Keywords: computer-aided drug design; virtual screening; leucyl t-RNA synthetase inhibition; molecular dynamics simulation

1. Introduction

The WHO Global Tuberculosis Report 2022 estimated that there are about 10.6 million cases of tuberculosis globally, along with 1.6 million deaths [1]. Incidences of drug-resistant tuberculosis have become common and are rising at an alarming rate. The development of resistance and cross-resistance has always driven researchers to search for novel targets and drugs with new mechanisms to counter the drug resistance problem [2,3]. Aminoacyl t-RNA synthetases are ubiquitously found in all organisms and regulate the transcription process by playing an active role in protein biosynthesis. It involves condensation of amino acid with ATP followed by transfer of the aminoacyl group to the 3′ end of homologous t-RNA. Aminoacyl t-RNA synthetases also regulate the mitochondrial RNA cleavage and proofreading of matured t-RNA [4]. High selectivity against bacteria can be achieved because of notable structural differences observed between prokaryotic and eukaryotic aminoacyl t-RNA synthetases in their catalytic domains [5]. Further, the aminoacyl t-RNA synthetases are not easily mutated, which confers it with a lower chance of developing drug resistance [4]. As per the active site topology features, aminoacyl t-RNA synthetases are divided into two types. The first class of enzymes contains a classical Rossmann dinucleotide-binding domain, whereas the second class of enzymes have anti-parallel β-sheets [6]. Leucyl t-RNA synthetase (LeuRS) is one of the first classes of aminoacyl t-RNA synthetase enzymes and has recently been validated as a promising antimycobacterial target. Leucyl t-RNA synthetase contains three different types of domains—an aminoacyl domain, a t-RNA binding domain, and an editing domain (also known as connective peptide) [7]. This editing domain has proofreading action and can hydrolyze any mischarged t-RNAs complexed with wrong aminoacyl groups [8]. Various benzoxaborole derivatives, like AN2690, ZCL039, GSK656, and EC/11770, have been discovered as potent inhibitors, amongst which GSK656 and EC/11770 possess antimycobacterial leucyl t-RNA synthetase [4,8] inhibition properties [9,10]. Some antimycobacterial leucyl t-RNA synthetase inhibitors amongst the N-Benzylidene-thiazol-2-yl-hydrazin [1] and 5-phenylamino-2H-[2,4]triazin-3-one derivatives have also been reported [11,12]. Figure 1 represents the known inhibitors against mycobacterial leucyl t-RNA synthetase.

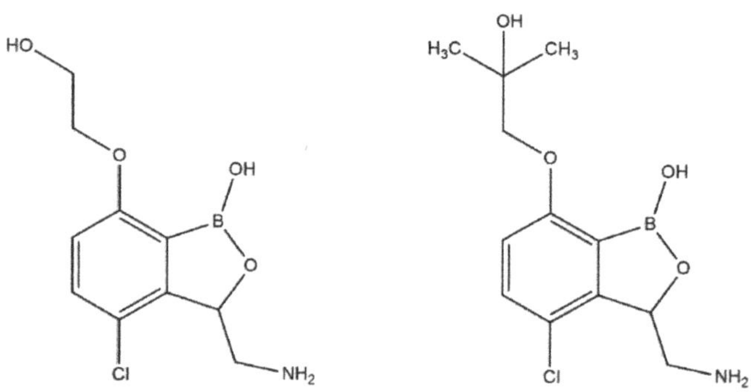

GSK656 **EC/11770**

Figure 1. Inhibitors of mycobacterial leucyl t-RNA synthetase.

Researchers are searching for lead molecules against new targets like aminoacyl t-RNA synthetase to prevent the looming threat of drug-resistant tuberculosis. Various computational drug design tools coupled with in silico screening have become an indispensable part of the current drug discovery programs [2]. Molecular simulation studies have shown high efficacy in predicting hit molecules that can be developed into novel drug candidates against leucyl t-RNA synthetase. Virtual screening has been extensively used to screen large databases ranging from small molecules to approved drug molecules to large compounds, both synthetic as well as natural, with diverse chemical structures [13,14]. A database consisting of drug molecules officially approved by various countries' regulatory authorities was screened. The molecular docking score has been routinely used in structure-based virtual screening to filter the databases. But often, it has been observed that compounds with good docking scores do not reside at the catalytic pocket of the receptor in the aqueous environment of the body. To address this problem, a molecular dynamics simulation study was conducted to find the best candidates for forming stable complexes with the protein. Finally, the top hit candidates were also assessed for their suitable pharmacokinetic and toxicity profiles. In this study, we have employed docking, MMGBSA energy analysis, and dynamics simulation modules of the Schrodinger suite to filter out the top hits against leucyl t-RNA synthetase from the prepared database. A biophysical method and an isothermal titration calorimetry (ITC) were used to study the binding affinity of the hit molecules. Biological activities like inhibition of biofilm and LeuRS gene expression studies were carried out to validate our in silico findings. This drug-repurposing study aims to identify drug molecules that have already been approved by the FDA and various other drug regulatory authorities. The selected drug molecule has the advantages of having known physico-chemical properties, clinical trial data, adverse effects, and safety profiles.

Therefore, if these in silico hits are to be selected for clinical trials against tuberculosis, then phase-I and phase-II studies could be skipped as the data are already available. Through structure-based virtual screening, we have identified three in silico hits that were found to have the potential to bind with the leucyl t-RNA synthetase of *Mycobacterium tuberculosis*. To validate the virtual screening results, the in silico hits were subjected to MD simulation, isothermal titration calorimetry, in vitro antimycobacterial activity assays, and tests for biofilm inhibition and LeuRS gene expression. A detailed workflow describing the present virtual screening campaign has been illustrated in Figure 2.

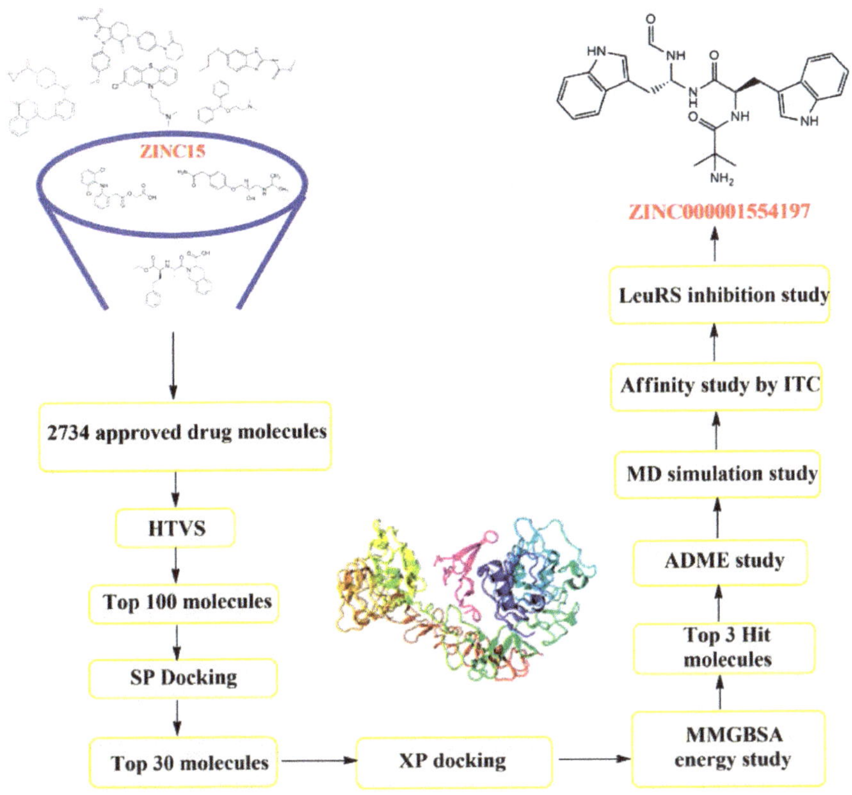

Figure 2. Detailed workflow of the virtual screening.

2. Materials and Methods

Schrodinger suite (2021-22 release) was used to execute this drug discovery endeavor.

2.1. Preparing Protein

The 5AGS.pdb crystal structure of leucyl t-RNA synthetase complexed with 3-(Aminomethyl)-4-bromo-7-ethoxybenzo[c][1,2]oxaborol-1(3H)-ol-adenosine was downloaded from protein data bank. The leucyl t-RNA synthetase crystal structure of *Mycobacterium tuberculosis* was determined by X-ray diffraction method with a resolution of 1.47 Å. Various amino acid residues like Thr336, Leu432, Tyr435, Asp447, Asp450, Met441, and Arg449 are some of the important binding residues that participate in interaction with the co-crystal ligand. This protein was processed, and a grid was generated using "protein preparation wizard" and "receptor grid generation" panel, as per the standard protocol [15,16].

2.2. Preparing Database

ZINC15 is an open-source database with thousands of approved drug molecules and drug-like compounds [17]. From this database, 2913 drug molecules were downloaded to prepare an in-house database. All the structures were cleaned up, and the database was screened according to the ring size, number of heteroatoms, and torsional centers, using LigPrep module to yield final 2734 approved drug molecules [18,19].

2.3. Structure-Based Virtual Screening

Glide module of Schrodinger suite has 3 modes of docking: HTVS (high-throughput virtual screening), SP (standard precision) docking, and XP (extra precision) docking. HTVS mode of docking with low accuracy was utilized for initial fast screening of the prepared database against 5AGS protein. Based on the HTVS scores, top 100 molecules were put forward for SP docking.

2.4. Glide XP Docking

The top 30 molecules obtained from the SP docking were again docked by using extra precision mode of Glide module using standard procedure [20–22].

2.5. MMGBSA

Prime module of Schrodinger suite was utilized for calculating MMGBSA energy of the docked complex [18].

2.6. ADMET Studies

Prior ADMET studies are essential to avoid late-stage failure of drug candidates in clinical trials. QikProp module of the Schrodinger suite was utilized to carry out prediction of pharmacokinetic parameters of top hit compounds [23,24].

2.7. Dynamics Simulation Studies

Molecular dynamics simulations for each protein–ligand complex were performed using the Desmond-maestro software package (v5.2) [18,25]. The system for each complex was built up by selecting TIP4P as a solvent model and orthorhombic as a box shape. Buffer method was used to calculate the box size, and the complexes were reoriented to minimize the box volume. Salt and ion placement were excluded within 20 Å. The structure was neutralized by adding sodium as a counter-ion. A default concentration of sodium chloride salt (0.15) was also set to start the system-building process. After system building, the energy minimization was performed using the energy minimization tool in Desmond-maestro at 100 ps. Each minimized system was exported for the production run in the MD simulation tool of the same software package. Further, NPT ensemble class and other default parameters were selected to start the 100 ns molecular dynamics simulation for each system. Obtained results were analyzed using the simulation interaction diagram module of the Desmond Maestro package [26,27].

2.8. Isothermal Calorimetry (ITC)

Isothermal titration calorimetry was used to examine the affinity between leucyl t-RNA synthetase (LeuRS) and ligands (compounds **1035**, **1054**, and **2077**). For ITC measurement, LeuRS and ligands (compounds **1035**, **1054**, and **2077**) were procured from Sigma-Aldrich at highest grade available. Isothermal titration calorimetric measurements were conducted with Malvern Instruments (Malvern, UK) at 25 °C and at pH 7.0 (ITC buffer), as described earlier [28]. LeuRS and ligands (compounds **1035**, **1054**, and **2077**) were dissolved in the ITC buffer 5 mM HEPES-KOH, pH 7.0, and the concentration of protein and ligands was determined spectrophotometrically using standard reference extinction coefficients. ITC experiments and measurements for LeuRS with the potential ligands (compounds **1035**, **1054**, and **2077**) were conducted separately, where ligands were allowed to bind with multiple binding sites. In this study, the top three hits (compounds **1035**, **1054**, and **2077**) were subjected to ITC measurements.

2.9. Inhibition of Mycobacterial Biofilm Formation

Wild-type *Mycobacterium tuberculosis* was grown in rich LB media overnight. The cell culture was diluted 1:1000 for biofilm inhibition assay. A routine biofilm assay medium was used with added magnesium sulfate, glucose, and casamino acid that support biofilm formation. Here, 200 µL of wild-type *Mycobacterium tuberculosis* culture + 100 µL biofilm assay medium was added and allowed to be incubated for 8 hr at 37 °C. After incubation, 100 µL crystal violet (0.1% crystal violet) was added to the tube and allowed for 20 min incubation at room temperature. Thereafter, 100 µL acetic acid (25% v/v acetic acid) was added to each tube and incubated for 15 min at room temperature. The tubes were subjected to absorbance readings at 550 nm. The test compounds 25 µL (10 mM) were added and allowed to be incubated for 30 mins, and absorbance was recorded at 600 nm. A positive control using standard benzoxaborole AN2690 as Leucyl-tRNA [8] synthetase was used [29,30].

2.10. Antimycobacterial Activity

Leucyl-tRNA synthetase (LeuRS) inhibition was achieved via compounds **1035**, **1054**, and **2077**, along with the determination of minimum inhibitory concentration. Here, in the study of the MIC determination, *Mycobacterium tuberculosis* strain H37Rv (*ATCC* 25618) was used. The screened ligands' (compounds **1035**, **1054**, and **2077**) anti-bacterial activity against the ATCC strain, *Mycobacterium tuberculosis* strain H37Rv (*ATCC* 25618), was determined by well diffusion method described previously [31]. The *Mycobacterium tuberculosis* strain H37Rv (*ATCC* 25618) was grown in the Löwenstein–Jensen medium (LJM) as per the recommended growth conditions. The stock solutions of ligands (compounds **1035**, **1054**, and **2077**) were prepared at a concentration of 10 mg/mL in the DMSO solution. Stock solution was further diluted to 1 mL/mL for preliminary screening. For the determination of minimum inhibitory concentration (MIC) of screened ligands (compounds **1035**, **1054**, and **2077**), the ligands were subjected to the serial dilution using DMSO solution at 6.25, 12.5, 25, 50, 100, 200, and 400 µg/mL. *Mycobacterium tuberculosis* strain H37Rv (*ATCC* 25618) was spread on the LJM agar medium. The wells were prepared on LJM agar plate, and 20 µL of each compound in a different concentration was added to the well and allowed to grow for 6 weeks. The zone of inhibition was measured, and MIC was determined.

2.11. Leucyl t-RNA Synthetase Gene Expression Inhibition Study

Hypotriploid alveolar basal epithelial cells A459 were used to examine the inhibitory effect of screened compounds on Leucyl-tRNA synthetase mRNA. Hypotriploid alveolar basal epithelial cells A549 were raised in Dulbecco's modified Eagle's medium nutrient mixture F-12 Ham (DMEM/F12) at 37 °C in a humidified 5% CO_2 incubator with supplements of 9% fetal bovine serum, 1% penicillin/streptomycin solution, and 1% L-glutamine. The cells were incubated with screened compounds by adding 25 µL (10 mM) and allowed to rest overnight. Total RNA of overnight grown hypotriploid alveolar basal epithelial cells A459 was isolated. Cells were washed with PBS, and qPCR analyses were carried out in triplicate using a novel fluorescent DNA-binding dye (like SyberGreen I) and an Applied Biosystems 7500 Fast Real-Time PCR System. The primers used here are forward primer (5′-TAATACGACTCACTATAGTCAGGATGGCCGAGCGGTCTA-3′) and reverse primer (5′-TGGTGTCAGGAGTGGGATTCGAACCCAC-3′). The copies of Leucyl-tRNA synthetase gene mRNA were examined per cycle with and without treatment of test compounds. Here, cytosolic Leucyl-tRNA synthetase gene inhibition was examined using test compounds along with a positive control (benzoxaborole AN2690) [9,32,33].

3. Results and Discussion

The ZINC15 database was used to prepare an in-house database containing 2913 approved drug molecules by the US FDA and other regulatory authorities across the world. This database was filtered and cleaned up using the LigPrep module to obtain the final 2734 compounds. The crystal structure of leucyl t-RNA synthetase complexed with 3-(Aminomethyl)-4-bromo-7-ethoxybenzo[c][1,2]oxaborol-1(3H)-ol-adenosine was utilized for screening the prepared database. Molecular dynamics simulation, ADMET, and MMGBSA studies were carried out on the top three hits.

3.1. Virtual Screening

The in-house database, which obtained 2734 compounds, was initially screened using the HTVS option of the Glide docking module. HTVS is a high-throughput virtual screening that takes less time to screen a large database by reducing the number of conformations and sampling space. The top 100 molecules obtained from the HTVS study were subjected to SP docking, which uses the same algorithm as that of HTVS but with standard precision. Based on the SP docking score of the compounds, the top 30 molecules were selected for further studies. The co-crystallized ligand of the 5AGS protein structure and a standard inhibitor GSK656, which is under clinical trial, has also been included in this virtual screening for a comparative study. The docking scores ranged from −11.172 to −2.556 Kcal/mol, whereas docking scores of the co-crystallized ligand and GSK656 were −6.69 and −6.508 Kcal/mol, respectively. The SP docking scores of the top 30 compounds, along with their structures, are presented in Supplementary Table S1.

3.2. Docking and MMGBSA Studies

The top 30 compounds selected from SP docking were further subjected to XP docking and MMGBSA energy analysis. Although XP docking uses a combination of scoring functions with enhanced accuracy and penalty for any violations, it does not take solvation energy and entropy factors into consideration. Therefore, the final selection of the top three hit molecules should be made based on the free energy values of the docked complexes. The prime module of the Schrodinger suite was utilized to determine the accurate free energy binding analysis of the docked complexes.

The MMGBSA energy values ranged from −87.56 to −39.73 Kcal/mol, whereas the standard inhibitor (GSK656) and the co-crystal ligand had MMGBSA values of −69.72 and −54.81 Kcal/mol, respectively. The XP docking scores and MMGBSA energy values of the top 30 molecules are presented in Table 1. All the compounds had a common binding site inside the 5AGS crystal structure of leucyl t-RNA synthetase. Compounds **1035** (ZINC000001543916), **1054** (ZINC000001554197), and **2077** (ZINC000008214483) were selected to be top hit compounds according to their MMGBSA energy values, which are −87.56, −79.79, and −77.34 Kcal/mol, respectively.

Table 1. Top 30 compounds with their XP docking scores and MMGBSA energy values.

Ligand	XP Docking Score	MMGBSA dG Bind
1035	−8.819	−87.56
1054	−9.006	−79.79
2077	−7.986	−77.34
1425	−6.908	−75.21
2350	−7.584	−74.15
1098	−6.394	−73.03
2102	−6.326	−73.02
1008	−7.245	−70.52
1256	−6.001	−67.01

Table 1. *Cont.*

Ligand	XP Docking Score	MMGBSA dG Bind
1104	−7.974	−61.83
179	−7.012	−59.99
365	−6.883	−58.99
207	−6.864	−58.95
906	−7.006	−57.78
2057	−5.738	−57.09
1327	−6.689	−55.11
97	−5.677	−54.91
2128	−6.641	−53.97
2103	−7.454	−52.49
2202	−7.385	−52.22
479	−6.294	−52.11
233	−6.176	−52.08
534	−6.131	−52.01
1033	−4.888	−51.05
2031	−4.852	−49.67
194	−5.582	−48.17
640	−4.598	−48.11
1213	−5.354	−46.84
118	−4.767	−45.71
2524	−3.091	−42.38
1973	−2.061	−39.73

3.3. Glide XP Docking

Various docking interactions of the top three hit compounds were compared to those of the co-crystallized ligand. The co-crystal ligand of 5AGS was re-docked to the binding cavity of the leucyl t-RNA synthetase. The compound bound to its previous attachment site and achieved an XP docking score of −6.206 Kcal/mol and MMGBSA binding free energy of −54.81 Kcal/mol. The oxaborole and adenosine moiety fitted inside the binding cavity, which was lined with residues like Thr336, Thr337, Leu432, Tyr435, Met441, Asp447, Arg449, and Asp450. The amino group of the adenosine moiety formed a hydrogen bond with the Leu432 residue, whereas the pyrimidine nucleus of the adenosine moiety showed a pi–pi hydrophobic interaction with the Try435 residue. Thr336 residue was involved in hydrogen bonding with the oxygen atom of the methoxy bridge connecting the oxaborole nucleus. Another amino group adjacent to the oxaborole nucleus was also involved in hydrogen bonding with Asp447 and Asp450 residues. The docking interactions of the co-crystal ligand are depicted in Figure 3.

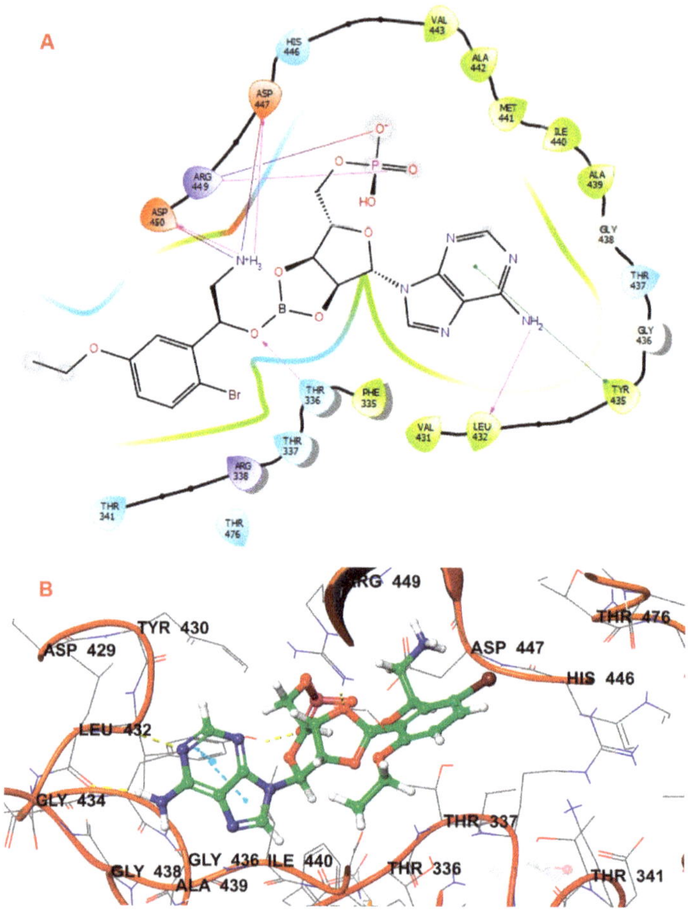

Figure 3. Interactions of the co-crystal ligand inside leucyl t-RNA synthetase. (**A**) 2D pose; (**B**) 3D pose.

Compound **1035** obtained an XP docking score of −8.819 Kcal/mol and an MMGBSA value of −87.56 Kcal/mol. The purine nucleus was buried inside the binding site, which was stabilized by the pi–pi interaction with the hydrophobic Tyr435 residue. A hydrogen bond was found between the fourth position carbonyl group attached to purine and Leu432 residue. Met441 and Asp450 residues were involved in hydrogen bonds with the amino group attached to the methyl butanoate chain. The ligand was further stabilized inside the binding cavity through hydrogen bonds involving Thr336 with the carbonyl group and Arg449 with the hydroxyl group. The docking interactions of the compound **1035** are depicted in Figure 4.

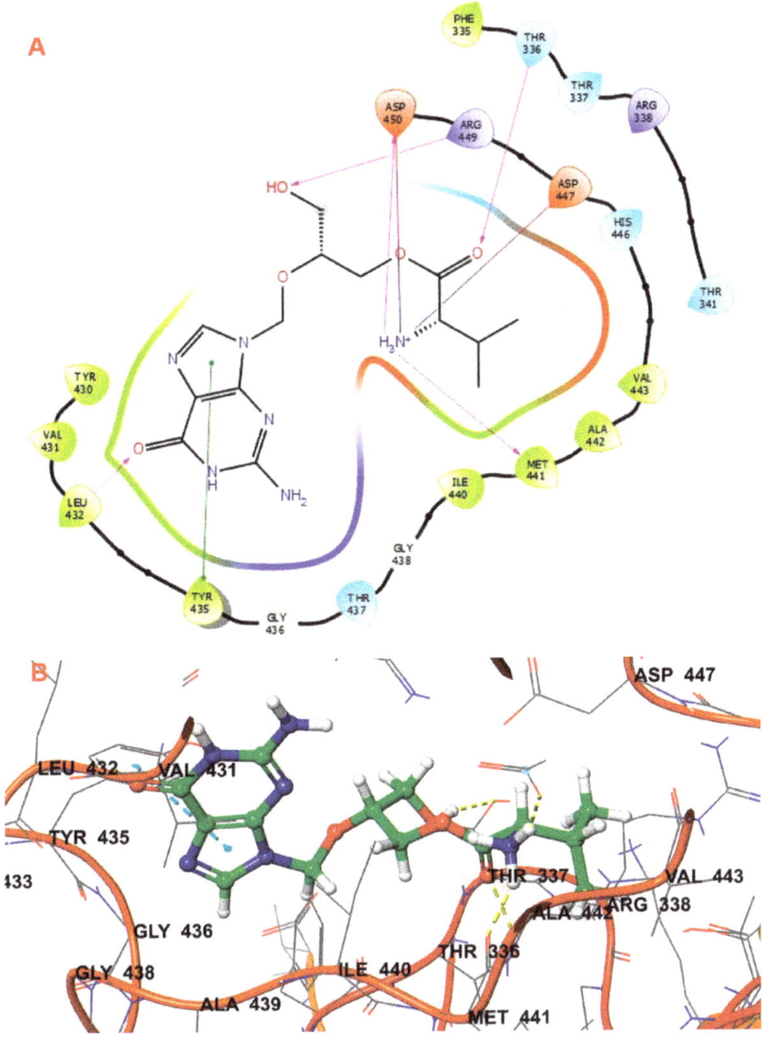

Figure 4. Interactions of 1035 inside leucyl t-RNA synthetase. (**A**) 2D pose; (**B**) 3D pose.

Compound **1054** obtained an XP docking score of −9.006 Kcal/mol and an MMGBSA value of −79.79 Kcal/mol. The indole nucleus attached to the formyl amino branch was tightly fitted inside the binding cavity, stabilized by the pi–pi interaction with the hydrophobic residue Tyr435. The heterocyclic nitrogen atom of the indole moiety also formed a hydrogen bond with Tyr435. Various carbonyl groups were involved in hydrogen bonding with residues Thr336, Thr337, and Arg449. The nitrogen of the indole moiety present at the opposite end formed a hydrogen bond with His446 residue. The terminal amino group was stabilized by hydrogen bonding with Asp447 and Asp450. The amino group attached to the methyl propanamide chain formed a hydrogen bond with residue Asp447. The docking interactions of the compound **1054** are depicted in Figure 5.

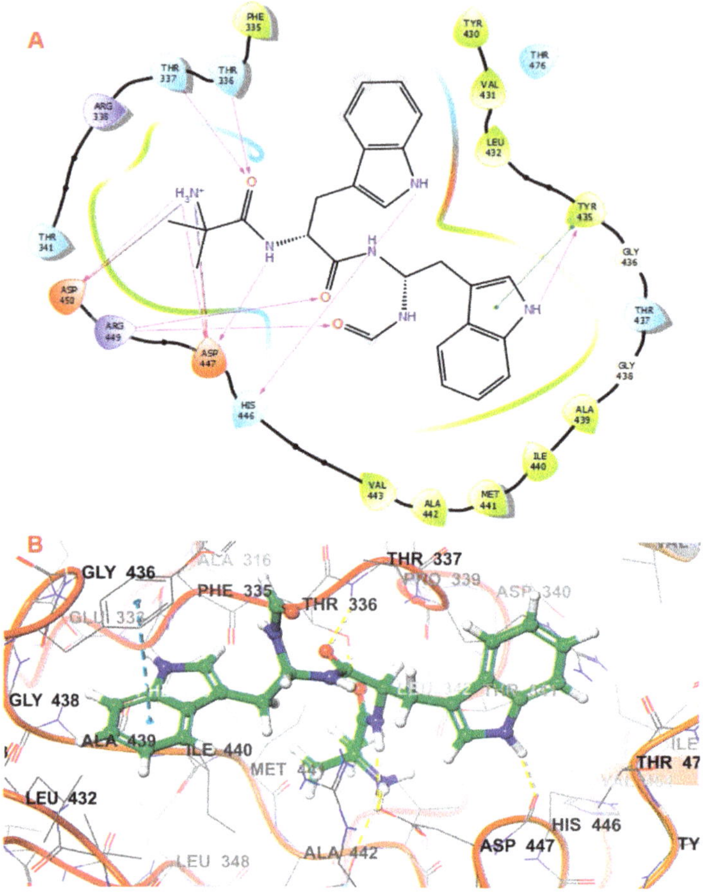

Figure 5. Interactions of 1054 inside leucyl t-RNA synthetase. (**A**) 2D pose; (**B**) 3D pose.

Compound **2077** obtained an XP docking score of −7.986 Kcal/mol and an MMGBSA value of −77.34 Kcal/mol. It is an aminoglycoside where two pyran rings are attached to a cyclohexane through glycosidic linkages. One of the pyran nuclei is stabilized inside the binding site through two hydrogen bonds between Tyr435 and Ala439 with the amino group and another hydrogen bond between Leu432 and the hydroxyl group. The terminal amino group was involved in hydrogen bonds with Asp447 and Asp450 residues. The hydroxyl group attached to the butanamide chain formed a hydrogen bond with residue Met441. The carbonyl group of the butanamide chain also helped in stabilizing the ligand by forming two hydrogen bonds with Thr336 and Thr337 residues. The amino group attached to the cyclohexane skeleton was involved in a hydrogen bond with Thr337 residue. The docking interactions of the compound **2077** are depicted in Figure 6.

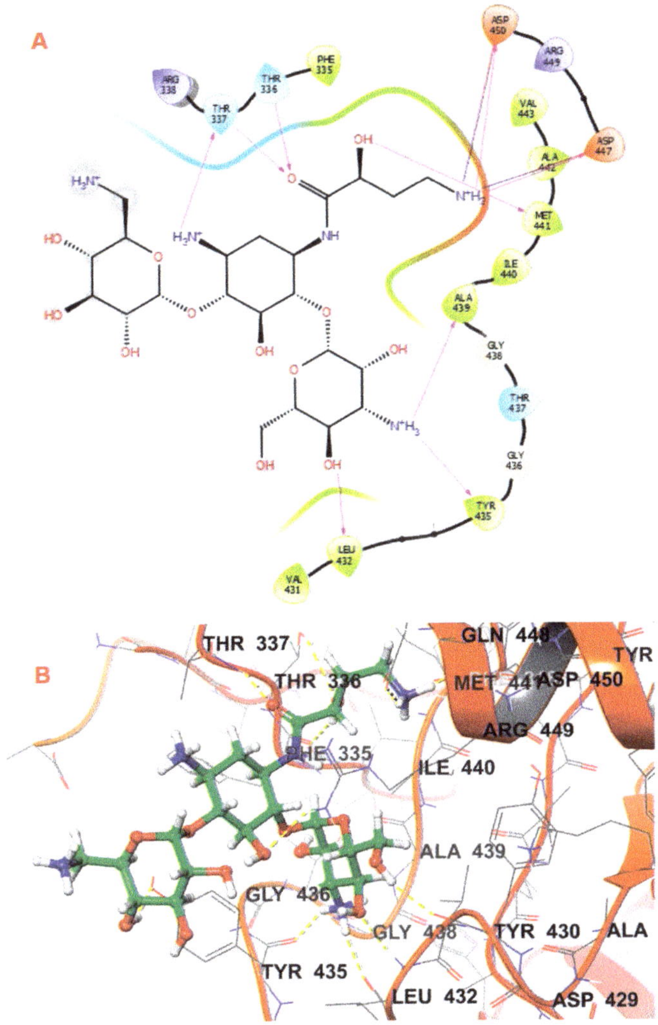

Figure 6. Interactions of 2077 inside leucyl t-RNA synthetase. (**A**) 2D pose; (**B**) 3D pose.

3.4. MMGBSA Studies

MMGBSA dG Bind is the summation of individual energy contributions due to coulomb energy, covalent binding energy, hydrogen bonding correction, lipophilic energy, pi–pi packing correction, self-contact correction, generalized Born electrostatic solvation energy, and van der Waals energy [34]. Based on the MMGBSA dG Bind values, compounds **1035** (ZINC000001543916), **1054** (ZINC000001554197), and **2077** (ZINC000008214483) were selected as the top three hit molecules. Contribution from MMGBSA dG Bind vdW was found to be maximum, whereas zero contribution was attributed from MMGBSA dG Bind SelfCont. In compound **1035**, the significant contributions for MMGBSA dG Bind were from MMGBSA dG Bind Lipo (−30.54 Kcal/mol), MMGBSA dG Bind Solv GB (33.41 Kcal/mol), and MMGBSA dG Bind vdW (−79.79 Kcal/mol). Similarly, for compound **1054**, important contributions were from MMGBSA dG Bind Lipo (−34.72 Kcal/mol), MMGBSA dG Bind Coulomb (29.14 Kcal/mol), and MMGBSA dG Bind vdW (−69.36 Kcal/mol). Likewise, contributions from MMGBSA dG Bind Lipo (−28.36 Kcal/mol), MMGBSA dG Bind Solv

GB (30.81 Kcal/mol), and MMGBSA dG Bind vdW (−66.42 Kcal/mol) were significant for compound **1077**.

3.5. Predicting Pharmacokinetic Parameters

A good drug candidate must have favorable pharmacokinetic parameters. As we have screened a database consisting of approved drug molecules, they can be assumed to have suitable pharmacokinetic parameters. QikProp module was used to evaluate the pharmacokinetic parameters of the top three hits, which are presented in Table 2. The molecular weights of compounds **1035** and **1054** were below 500 g/mol, whereas compound **2077** has a molecular weight of 585.607 g/mol. The dipole moment, SASA (total solvent accessible surface area), and volume (total solvent accessible volume) for all three compounds were found to be within the range (i.e., 1 to 12,300 to 100 Å2 and 500 to 2000 Å3, respectively). The van der Waal's surface area due to polar N, O, and carbonyl C were within the range for compounds **1035** and **1054**, whereas it crossed the upper limit in the case of compound **2077**. Compounds **1035** and **1054** have suitable values for the predicted octanol–water partition coefficient (QPlogPo/w), predicted brain–blood partition coefficient (QPlogBB), predicted skin permeability (QPlogKp), and predicted binding to human serum albumin (QPlogKhsa), but for 2077, the values lie outside the approved range. Except for compound **2077**, the predicted aqueous solubility (QPlogS) values were found to be within the range. All three compounds had poor transport across the gut–blood barrier, as predicted from QPPCaco values. The predicted oral absorption was poor for compounds **1035** and **2077**, but medium absorption was predicted for compound **1054**. The number of violations, as per the rule of five and the rule of three, was within the recommended range for all three compounds.

Table 2. Pharmacokinetic parameters.

Ligand	MW	Dipole	SASA	PSA	Volume	QPlog Po/w	QPlogS	QPP Caco	QPlog BB	QPlog Kp	QPlog Khsa	Oral Absorption	Rule of Five	Rule of Three
1035	354.365	10.74	623.453	176.713	1098.301	−1.464	−1.054	5.248	−2.457	−7.497	−0.925	1	2	1
1054	474.561	6.49	791.223	156.869	1462.906	2.07	−3.236	10.173	−1.824	−5.008	−0.199	2	1	1
2077	585.607	11.953	859.859	322.444	1629.256	−8.741	−0.194	0.002	−5.453	−12.732	−2.65	1	3	2

MW: molecular weight; SASA: total solvent accessible surface area; PSA: polar surface area; QPlogPo/w: octanol–water partition coefficient; QPlogS: aqueous solubility; QPPCaco: Caco-2 cell permeability; QPlogBB: brain–blood partition coefficient; QPlogKp: skin permeability; QPlogKhsa: binding to serum albumin protein.

3.6. Dynamics Simulation Studies

Based on the XP docking and MMGBSA energy values, the top three hits were chosen (presented in Table 3) for molecular dynamic (MD) simulation to analyze the structural behavior of the protein–ligand complex. The docked complex stability was assessed by MD simulations for 100ns of the protein–ligand complexes. The changes in the complex conformations were evaluated by the root mean square deviation (RMSD) and root mean square fluctuation (RMSF) values.

Table 3. Top three hit molecules.

Ligand	Structure	IUPAC Name	XP Docking Score	MMGBSA dG Bind
1035 (ZINC000001543916) Valganciclovir	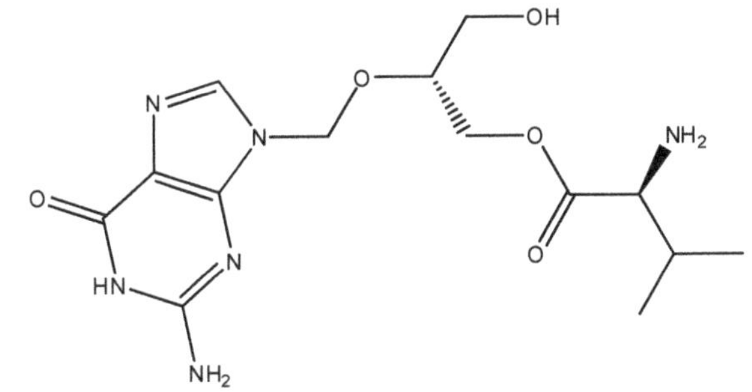	2-((2-amino-6-oxo-1H-purin-9(6H)-yl)methoxy)-3-hydroxypropyl 2-amino-3-methylbutanoate	−8.819	−87.56
1054 (ZINC000001554197) Macimorelin	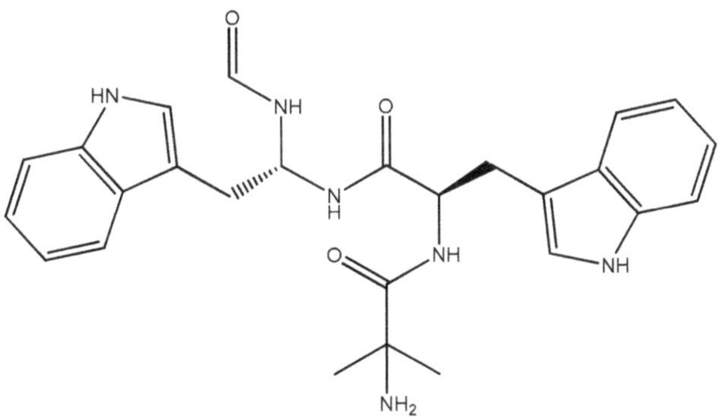	2-amino-N-((R)-1-(((R)-1-formamido-2-(1H-indol-3-yl)ethyl)amino)-3-(1H-indol-3-yl)-1-oxopropan-2-yl)-2-methylpropanamide	−9.006	−79.79
2077 (ZINC000008214483) Amikacin	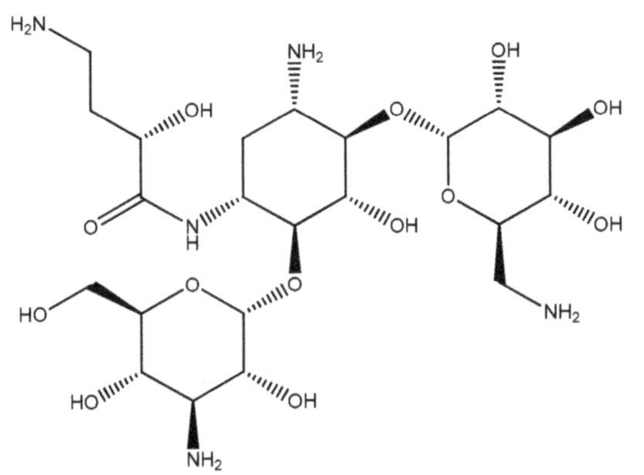	(S)-4-amino-N-((1R,2S,3S,4R,5S)-5-amino-2-(((2S,3R,4S,5S,6R)-4-amino-3,5-dihydroxy-6-(hydroxymethyl)tetrahydro-2H-pyran-2-yl)oxy)-4-(((2R,3R,4S,5S,6R)-6-(aminomethyl)-3,4,5-trihydroxytetrahydro-2H-pyran-2-yl)oxy)-3-hydroxycyclohexyl)-2-hydroxybutanamide	−7.986	−77.34

The RMSD of protein of the 5AGS_1035 complex in the graph (Figure 7A) shows slight deviations with 1–2 Å, and after 85 ns, it has a gradual increase of 3 Å. Meanwhile, the ligand RMSD appears with an initial RMSD value of 2.5 Å, followed by an increase in RMSD to 4.5 Å; a fall at 55–60 ns has been observed up to 2.5 Å, which then deviated up to 6 Å, with a stable conformation that remains in the protein pocket. The RMSD plot of the 5AGS_1054 complex showed that the protein and ligand were stable and remained in the pocket with acceptable fluctuations during the simulation of 100 ns with a protein RMSD value of 1–2 Å and a ligand RMSD of 2–4 Å (Figure 7B). In contrast, the RMSD plot of the 5AGS_2077 complex shows an initial protein RMSD value between 1.2 and 2.8 Å, which shows less conformational variations, but the ligand RMSD started with small deviations until 20 ns, and then it shows an increase in the values to 30 Å at 25 ns, which again increased up to 60 Å between 75 and 85 ns. At the end of the 100 ns, the fluctuation decreased to 20 Å. Higher fluctuations in the RMSD plot suggested that the ligand might have come out of the binding pocket during the simulation. Therefore, it never reached the equilibrium and demonstrated unstable behavior of the ligand in the binding cavity of the protein (Figure 7C).

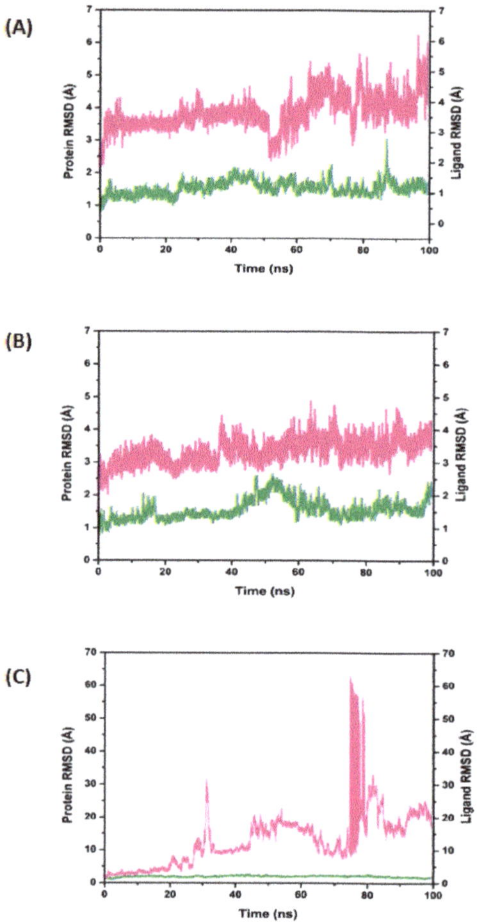

Figure 7. Root mean square deviation (RMSD) of the protein and ligand. Green color represents the target protein, whereas the magenta color represents the compounds, respectively: (**A**) 5AGS_1035 complex; (**B**) 5AGS_1054 complex; (**C**) 5AGS_2077 complex.

RMSF plots were also evaluated in the molecular dynamics simulation for each residue to determine their flexibility and conformational behavior. The protein RMSF values of these docked complexes were analyzed throughout the 100 ns dynamics simulation, which found that the 5AGS_1035 complex has fluctuation in the Cα of protein with ~4.5 Å (Figure 8A), while the 5AGS_1054 complex experienced this fluctuation until 4 Å (Figure 8B), and 5AGS_2077 complex showed fluctuations up to 3 Å (Figure 8C).

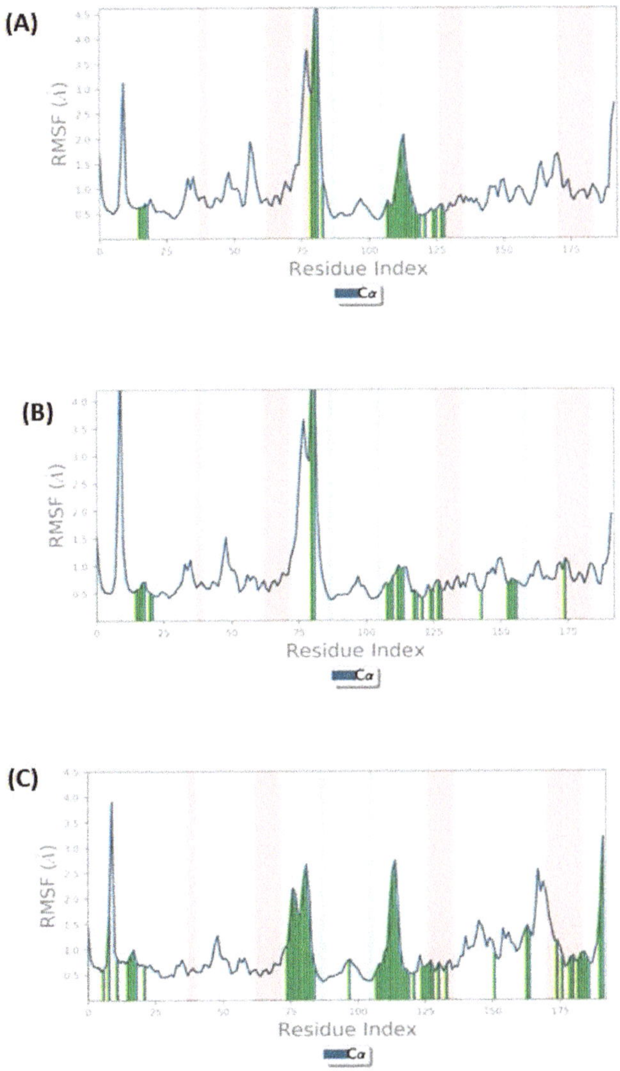

Figure 8. Protein root mean square fluctuation (P-RMSF) of docked complexes. (**A**) 5AGS_1035 complex; (**B**) 5AGS_1054 complex; (**C**) 5AGS_2077 complex. The blue line shows RMSF, the green line represents the interactions formed during simulation, the pink columns indicate alpha helices, and the blue columns indicate beta-strands.

The ligand RMSF values showed fluctuations within 1–3 Å in compound **1035** (Figure 9A), with more fluctuations in the C-terminal atoms. Similarly, compound **1054** showed fluctuation in atoms throughout the simulation ranging from 0.5–2 Å (Figure 9B), whereas a higher fluctuation range (above 10 Å) is observed in the case of 5AGS_2077 complex (Figure 9C).

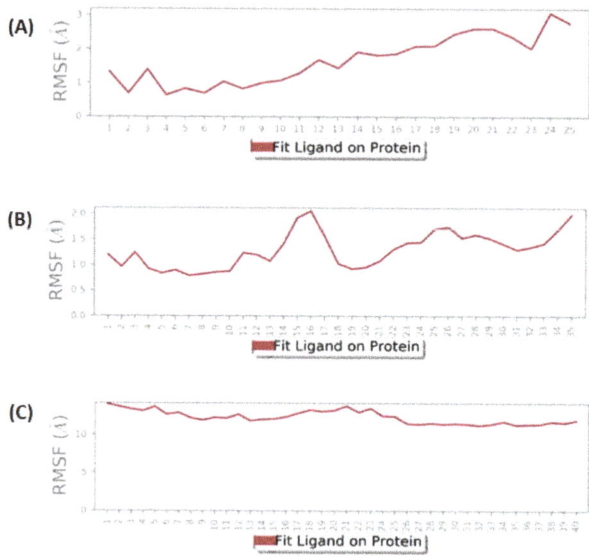

Figure 9. Ligand root mean square fluctuation (L-RMSF) of docked complexes. (**A**) 5AGS_1035 complex; (**B**) 5AGS_1054 complex; (**C**) 5AGS_2077 complex.

A protein–ligand mapping plot of the 5AGS_1035 complex (Figure 10A) revealed thatThr337, Leu432, Tyr435, Met441, Asp447, and Asp450 exhibited hydrogen bonding throughout the 100 ns simulation, while some of the residues also showed hydrophobic interactions (Ile440, Val443, Arg449), an ionic interaction in Asp447, and water bridges in some residues. However, the major residues in the ligand–protein contact reveal 39% and 72% contact over the span of the simulation in the Thr337 and Met441 residues, respectively, whereas the Asp450 residue exhibits more than 100% of the hydrogen bond interaction over the span of the entire simulation. The 2D interaction diagram (Figure 11A) revealed that the amino group (attached to the methyl butanoate chain) was involved in hydrogen bonding with Met441 and Asp450 residues. Another hydrogen bond was observed between the carbonyl group (of methyl butanoate chain) and Thr337 residue.

In the 5AGS_1054 complex, the hydrogen bond interactions involved residues like Thr336, Thr337, Arg338, His446, Asp447, and Asp450, whereas hydrophobic interactions involved Tyr435 and Arg449 residues with Thr337 residue accounting for more than 100% of the interaction time. Ligand protein contact (Figure 10B) indicates that Thr337 residue accounted for more than 100% of the interaction time; Tyr435 and Arg449 accounted for 35% and 46%, respectively, during the simulation. The interaction rates for other important residues like His446, Asp447, Asp450, and Thr336 were 80%, 59%, 39%, and 89%, respectively. The 2D interaction diagram (Figure 11B) showed hydrogen bonding with residues Thr336, Thr337, His446, Asp447, and Asp450; pi–pi contacts with Tyr435 and Arg449 residues; and water bridge interactions with Asp447 and Asp450 residues, respectively. The nitrogen of the indole nucleus was involved in a hydrogen bond with His446 residue. The terminal amino group was stabilized by hydrogen bonding with Asp447 and Asp450. The amino group attached to the methyl propanamide chain also formed a hydrogen bond with residue Asp447. The carbonyl groups of the methyl propenamide chain were involved in hydrogen bonding with Thr336 and Thr337 residues. The indole nucleus had shown pi–pi stacking interactions with Tyr435 and Arg449 residues.

In the 5AGS_2077 complex, most of the important residues showed less than 60% hydrogen bond interactions except for Glu401, which accounted for over 70% of the interactions (Figure 10C). No residues were stable in the ligand–protein contact, resulting in an unstable conformational shift in the 5AGS_2077 complex. Even the 2D interaction diagram did not generate any significant interactions. These results correlated well with the higher fluctuations observed in the RMSD plot of the 5AGS_2077 complex. It also confirms that the ligand moved away from the binding pocket during MD simulation and, therefore, lost most of the contacts. These results established the unstable nature of ligand 2077 inside the binding cavity of the protein.

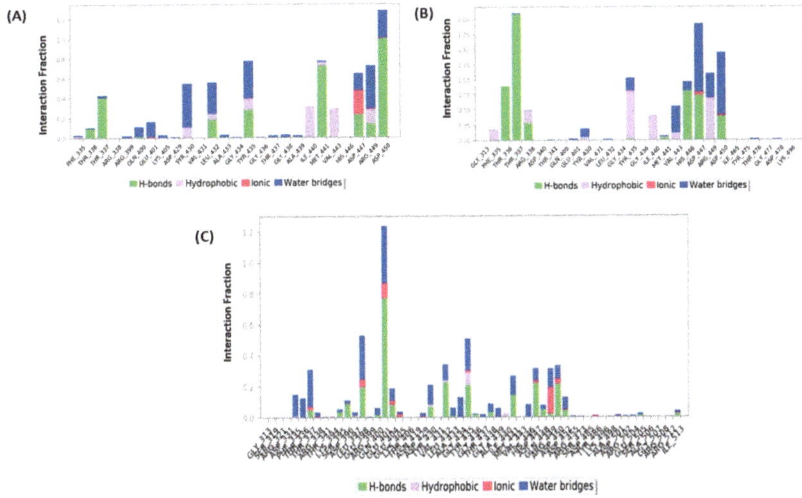

Figure 10. Protein–ligand contact mapping. (**A**) 5AGS_1035 complex; (**B**) 5AGS_1054 complex; (**C**) 5AGS_2077 complex.

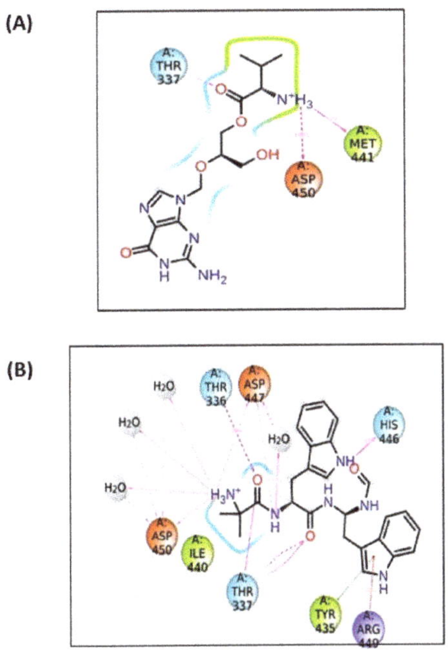

Figure 11. Two-dimensional representation of ligand–protein contacts. (**A**) 5AGS_1035 complex; (**B**) 5AGS_1054 complex.

3.7. Comparison of Interactions after XP Docking and Dynamics Simulation

The last pose of the MD simulation trajectory was compared with the XP docking results (presented in Table 4) to analyze the interactions that were retained in the hydrophilic environment during simulation.

Table 4. Interactions before and after MD simulation.

Compound	Interactions					
	After XP Docking			After MD Simulation		
	H Bonding	Pi–Pi	Salt Bridge	H Bonding	Pi–Pi	Water Bridge
1035	Thr336, Leu432, Met441, Arg449, Asp450	Tyr435	Asp447, Asp450	Thr337, Met441, Asp450	-	-
1054	Thr336, Thr337, Tyr435, His446, Asp447, Arg449, Asp450	Tyr435	Asp447, Asp450	Thr336, Thr337, His446, Asp447, Asp450	Tyr435, Arg449	Asp447, Asp450
2077	Thr336, Thr337, Leu432, Tyr435, Ala439, Met441, Asp447, Asp450	-	Asp447, Asp450	-	-	-

In compound **1035**, hydrogen bonds with Met441 and Asp450 residues were retained after dynamics simulation, whereas a new hydrogen bond with Thr337 was formed. No other pi–pi stacking and salt bridge interactions were retained during the dynamics study. In compound **1054**, hydrogen bonding interaction with residues Thr336, Thr337, His446, Asp447, and Asp450 were retained during dynamics simulation. Pi–pi stacking with Tyr435 residue was preserved, whereas a new pi–pi interaction with Arg449 residue was also produced. The salt bridge interactions observed in XP docking were retained in the form of water bridges through Asp447 and Asp450 residues. However, compound **2077** could not retain any of the interactions observed during XP docking.

On careful study of the docking, MMGBSA, and molecular dynamics results, compound **1054** or ZINC000001554197 (Macimorelin) was selected as the top hit compound. In the dynamics simulation, Macimorelin showed the maximum number of interactions with better stability (as evidenced by its RMSD and RMSF analysis) than the other two compounds. Macimorelin has two indole rings attached to each other by alkyl amidic linkage. The methyl propenamide substituent helped stabilize the compound inside the binding cavity through the formation of hydrogen bonds with water bridges. The indole nucleus was also helpful in providing pi–pi stacking interaction and involved with a hydrogen bond. Thus, the molecular dynamics study validated the initial results obtained from XP docking and MMGBSA studies.

3.8. Isothermal Calorimetry (ITC)

The top three hit molecules were tested for their binding affinity toward the LeuRS protein. Figure 12 demonstrates the course of reaction between the ligands and LeuRS protein until it reaches saturation. After about 105 min, active sites of the protein became completely saturated or occupied with the ligands, as observed by the absence of change in the peaks. As all three ligands have similar binding patterns, they also exhibited quite similar affinity. The results were fitted in the model, and binding parameters, i.e., the affinity between LeuRS and ligands (compounds **1035**, **1054**, and **2077**), were measured via the floating association constant (ka), binding enthalpy (ΔH), and the number of binding sites (n).

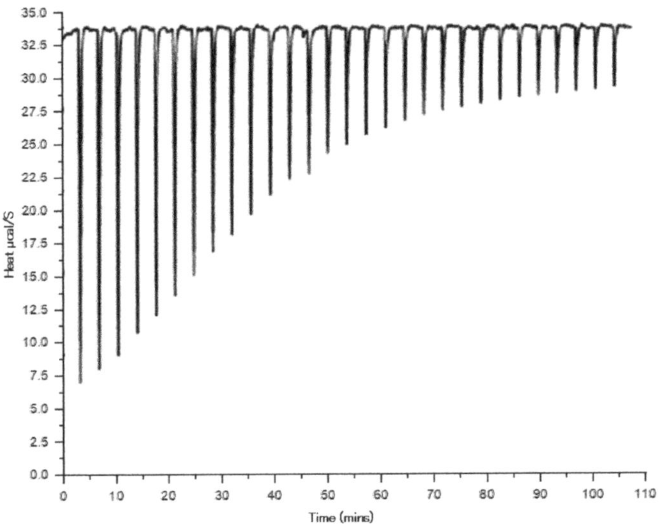

Figure 12. Course of reaction between the ligands and protein LeuRS.

The isothermal titration calorimetry measurement of the binding constant between LeuRS and the top three hits at 25 °C temperature and pH 7.0 were summarized in Table 5 and depicted in Figure 13.

Table 5. Constant binding between LeuRS protein and the ligands.

Conc. of Ligand/Protein (µM)	ΔH (kJ/mol) for Ligands		
	1035	1054	2077
00	−86.84	−82.11	−84.16
0.0125	−81.65	−76.15	−80.13
0.025	−75.56	−68.25	−73.79
0.050	−63.41	−60.11	−61.36
0.100	−57.85	−52.27	−55.16
0.200	−43.88	−41.25	−41.42
0.400	−34.16	−29.73	−31.25
0.800	−23.10	−20.46	−20.16
1.600	−16.25	−13.16	−13.45
3.20	−12.65	−10.42	−10.36
5.00	00	00	00

The data collected from ITC measurement clearly demonstrate that the affinity of LeuRS with compound **1054** was much higher than the other two ligands used in the present study, i.e., compounds **1035** and **2077**. The higher affinity of compound **1054** was in accordance with the affinity parameter determined via ITC, i.e., the floating association constant (Ka) was reported to be 6.32×10^6, binding enthalpy (ΔH) −48.23 kl/mol, and higher binding sites (n), i.e., 1.72. Table 6 summarizes the ITC results showcasing the floating association constant, binding enthalpy, and the number of binding sites. These findings support the in silico study findings used to test the affinity between LeuRS and other ligands. The affinity of compound **1054** or ZINC000001554197 (Macimorelin) was found to be superior over the other two ligands (compounds **1035** and **2077**), which corroborated the in silico findings. The other two ligands (compounds **1035** and **2077**) also showed promising affinity with the LeuRS and could be potential candidates for the drug development process.

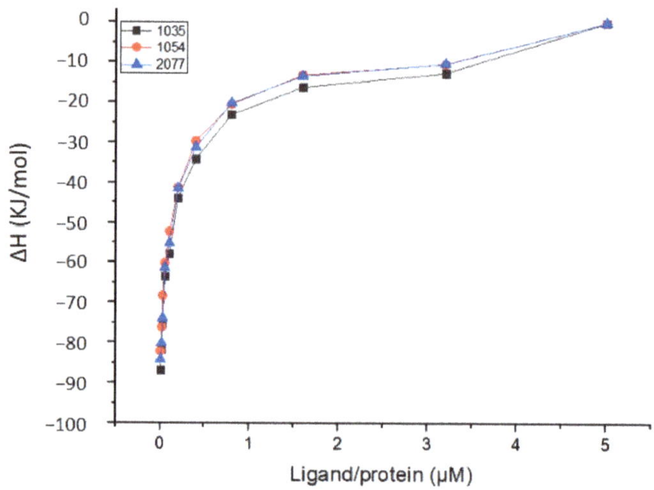

Figure 13. Binding affinity between LeuRS protein and compounds **1035**, **1054**, and **2077**.

Table 6. ITC measurements of compounds **1035**, **1054**, and **2077** against LeuRS.

Ligands	Floating Association Constant ka (1/M)	Binding Enthalpy ΔH (kJ/mol)	Binding Sites (n)
1035	4.18×10^4	−49.16	1.14
1054	6.32×10^6	−48.23	1.72
2077	7.74×10^3	−49.87	0.85

3.9. Inhibition of Mycobacterial Biofilm Formation

In the microbial community, biofilm formation is governed by a series of enzyme-catalyzed extracellular polymeric substances. Leucyl t-RNA synthetase is the key enzyme in biofilm formation in *Mycobacterium tuberculosis*, and it promotes growth and development [29]. Hence, inhibiting the leucyl t-RNA synthetase also affects the biofilm formation and attenuates the *Mycobacterium tuberculosis* growth. Here, in the present study, test compounds (**1035**, **1054**, and **2077**) were examined for biofilm formation inhibitory activity. All three compounds showed promising biofilm formation inhibitory activity as compared to control (data presented in Table 7). Figure 14 demonstrated that compound **1054** had shown higher inhibition in biofilm formation compared to the other two compounds. The findings of this study align with previous research indicating that inhibitors of Leucyl-tRNA synthetase also inhibit biofilm formation by regulating extracellular polymeric substance assembly [32].

Table 7. Biofilm formation inhibitory activity by compounds **1035**, **1054**, and **2077** at different concentrations with control (benzoxaborole AN2690).

2077		1035		1054		Control	
Conc.	OD	Conc.	OD	Conc.	OD	Conc.	OD
0	0.9195	0	0.8658	0	0.6804	0	0.5482
5.0	0.7588	5.0	0.7685	5.0	0.5428	5.0	0.4256
10.0	0.5358	10.0	0.6255	10.0	0.4548	10.0	0.3658
15.0	0.3585	15.0	0.5454	15.0	0.3425	15.0	0.2452
20.0	0.2544	20.0	0.3256	20.0	0.2186	20.0	0.2048
25.0	0.1456	25.0	0.1854	25.0	0.1254	25.0	0.0980

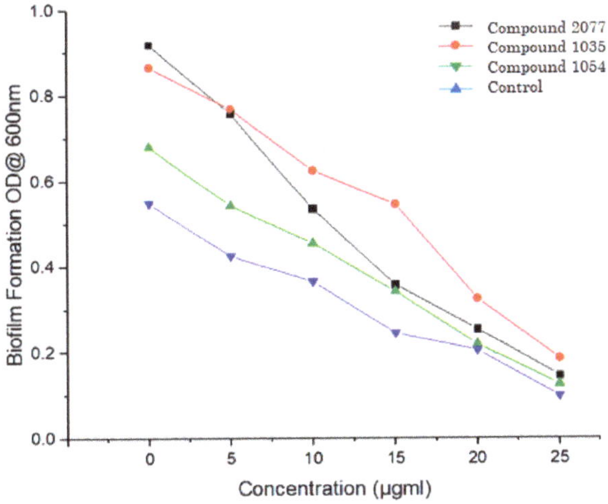

Figure 14. Inhibition of biofilm formation by compounds **1035**, **1054**, and **2077** at different concentrations with control (benzoxaborole AN2690).

Efforts have been continuously made to develop antimycobacterial drugs to combat the growth of *Mycobacterium tuberculosis*. During the drug development process, multiple targets within the MTB proteome have been identified, leading to variations in drug efficacy. In the present study, the antitubercular activity of ligands (compounds **1035**, **1054**, and **2077**) was examined by determining the zone of inhibition. As shown in Table 8, compound **1054** appears more promising compared to 1035 and 2077. The zone of inhibition (ZOI) for 1054 was measured at 18.45 ± 0.19 mm with a minimum inhibitory concentration (MIC) value of 186.39 ± 0.21 µg/mL. Although compounds **1035** and **2077** also exhibited antimycobacterial activity, compound **1054** demonstrated superior efficacy.

Table 8. Antimycobacterial activity of compounds **1035**, **1054**, and **2077** and minimum inhibitory concentration determination.

Ligands	**Anti-Mycobacteial Activity**	
	Zone of Inhibition (mm)	**MIC (µg/mL)**
1035	13.68 ± 0.21	119.87 ± 0.10
1054	18.45 ± 0.19	186.39 ± 0.21
2077	15.57 ± 0.80	147.43 ± 0.80

3.10. Leucyl t-RNA Synthetase Inhibition Study

Leucyl t-RNA synthetase is the key enzyme responsible for growth, development, and virulence [32]. Therefore, inhibiting this enzyme restricts the survival of the mycobacterial population and the management of tuberculosis. *Mycobacterium tuberculosis* possesses the LeuRS gene in its genome and mitochondria that encode the enzyme. Several studies have demonstrated that cytosolic LeuRS is more functional and effective in the growth and development of *Mycobacterium tuberculosis* [8]. The test compounds, along with the positive control AN2690, were evaluated for translation inhibitory role via inhibiting the LeuRS gene, where mRNA folds per cycle were determined. Certainly, benzoxaborole AN2690 showed lower copies of LeuRS in the RTPCR cycle, whereas compound **1054** was more effective in down-regulating the LeuRS gene as compared to the compounds **1035** and **2077** (Figure 15).

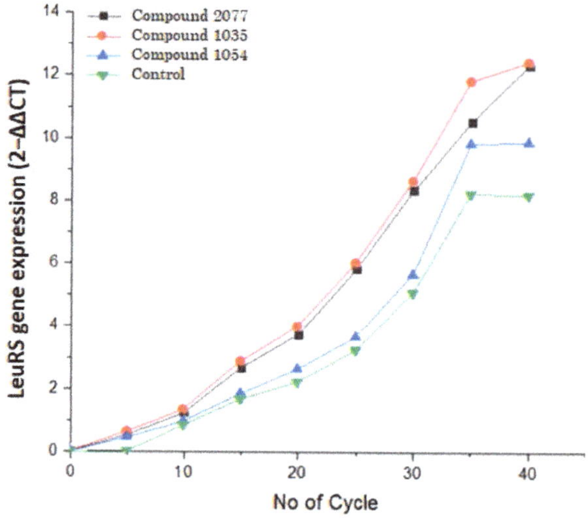

Figure 15. Inhibition of Leucyl-tRNA synthetase (LeuRS) gene expression; copies of mRNA per cycle under the influence of compounds **1035**, **1054**, and **2077**, along with control benzoxaborole AN2690.

4. Conclusions

A generated in-house database with 2734 drug molecules was screened against the leucyl t-RNA synthetase enzyme. Based on the docking scores and MMGBSA energy analysis, the top three hit molecules were selected. These hit molecules had a strong affinity toward the leucyl t-RNA synthetase active site, and their pharmacokinetic parameters were also found to be satisfactory. A molecular dynamics simulation study was carried out to ascertain the best hit molecule in terms of better stability and interactions inside the binding site. Compound **1054** or ZINC000001554197 (Macimorelin) showed comparatively better stability and retained many interactions observed during XP docking. Macimorelin was found to be the most potent among the three hit molecules. A biophysical study by ITC confirmed the superior affinity of compound **1054** toward LeuRS protein. Macimorelin showed significant antimycobacterial and mycobacterial biofilm inhibitory activity along with inhibition of LeuRS gene expression. Therefore, this in silico hit can be used as a lead compound for the development of more leucyl t-RNA synthetase inhibitors to target tuberculosis.

Supplementary Materials: The following supporting information can be downloaded at: https://www.mdpi.com/article/10.3390/biom14060711/s1, Table S1: Structure and docking scores of compounds.

Author Contributions: Conceptualization, Z.U.R. and A.N.; methodology, K.Z.; software, K.Z.; validation, Z.U.R., A.N. and K.Z.; formal analysis, A.N.; investigation, K.Z.; resources, A.N.; data curation, K.Z.; writing—original draft preparation, Z.U.R., A.N. and K.Z.; writing—review and editing, Z.U.R.; visualization, Z.U.R.; supervision, Z.U.R.; project administration, Z.U.R.; funding acquisition, A.N. All authors have read and agreed to the published version of the manuscript.

Funding: The authors gratefully acknowledge the funding of the Deanship of Graduate Studies and Scientific Research, Jazan University, Saudi Arabia, through Project Number: GSSRD-24.

Institutional Review Board Statement: Not applicable.

Informed Consent Statement: Not applicable.

Data Availability Statement: Data are contained within the article.

Conflicts of Interest: The authors declare no conflicts of interest.

References

1. Global Tuberculosis Report 2022. Available online: https://www.who.int/teams/global-tuberculosis-programme/tb-reports/global-tuberculosis-report-2022 (accessed on 20 October 2023).
2. Ndagi, U.; Kumalo, H.M.; Mhlongo, N.N. A Consequence of Drug Targeting of Aminoacyl-TRNA Synthetases in *Mycobacterium tuberculosis*. *Chem. Biol. Drug Des.* **2021**, *98*, 421–434. [CrossRef] [PubMed]
3. Faiz Arshad, M.; Kumar, S.; Al Rohaimi, A.H.; Hassan, A.A.; Elkerdasy, A.; Upmanyu, N. Synthesis of 3-Chloro-4-Fluorophenyl-3,4-Dichlorophenyl Substituted Thiocarbamide Derivatives as Potential Antitubercular Agents. *Med. Chem. Res.* **2015**, *24*, 334–343. [CrossRef]
4. Zhang, P.; Ma, S. Recent Development of Leucyl-TRNA Synthetase Inhibitors as Antimicrobial Agents. *Medchemcomm* **2019**, *10*, 1329–1341. [CrossRef] [PubMed]
5. Alsulaimany, F.A.; Almukadi, H.; Zabermawi, N.M.O.; Aljuhani, T.A.; Rashidi, O.M.; Albaqami, W.F.; Alghamdi, A.A.; Ahmad, A.; Shaik, N.A.; Banaganapalli, B. Identification of Novel *Mycobacterium tuberculosis* Leucyl-TRNA Synthetase Inhibitor Using a Knowledge-Based Computational Screening Approach. *J. King Saud Univ.—Sci.* **2022**, *34*, 102032. [CrossRef]
6. Hu, Q.H.; Huang, Q.; Wang, E.D. Crucial Role of the C-Terminal Domain of *Mycobacterium tuberculosis* Leucyl-TRNA Synthetase in Aminoacylation and Editing. *Nucleic Acids Res.* **2013**, *41*, 1859–1872. [CrossRef] [PubMed]
7. Liu, R.J.; Long, T.; Li, H.; Zhao, J.; Li, J.; Wang, M.; Palencia, A.; Lin, J.; Cusack, S.; Wang, E.D. Molecular Basis of the Multifaceted Functions of Human Leucyl-TRNA Synthetase in Protein Synthesis and Beyond. *Nucleic Acids Res.* **2020**, *48*, 4946–4959. [CrossRef] [PubMed]
8. Ganapathy, U.S.; González del Rio, R.; Cacho-Izquierdo, M.; Ortega, F.; Lelièvre, J.; Barros-Aguirre, D.; Lindman, M.; Dartois, V.; Gengenbacher, M.; Dick, T. A Leucyl-TRNA Synthetase Inhibitor with Broad-Spectrum Antimycobacterial Activity. *Antimicrob. Agents Chemother.* **2021**, *65*, 10–128. [CrossRef] [PubMed]
9. Palencia, A.; Li, X.; Bu, W.; Choi, W.; Ding, C.Z.; Easom, E.E.; Feng, L.; Hernandez, V.; Houston, P.; Liu, L.; et al. Discovery of Novel Oral Protein Synthesis Inhibitors of Mycobacterium tuberculosis that Target Leucyl-TRNA Synthetase. *Antimicrob. Agents Chemother.* **2016**, *60*, 6271–6280. [CrossRef] [PubMed]
10. Li, X.; Hernandez, V.; Rock, F.L.; Choi, W.; Mak, Y.S.L.; Mohan, M.; Mao, W.; Zhou, Y.; Easom, E.E.; Plattner, J.J.; et al. Discovery of a Potent and Specific M. Tuberculosis Leucyl-TRNA Synthetase Inhibitor: (S)-3-(Aminomethyl)-4-Chloro-7-(2-Hydroxyethoxy)Benzo[c][1,2]Oxaborol-1(3H)-Ol (GSK656). *J. Med. Chem.* **2017**, *60*, 8011–8026. [CrossRef] [PubMed]
11. Gudzera, O.I.; Golub, A.G.; Bdzhola, V.G.; Volynets, G.P.; Kovalenko, O.P.; Boyarshin, K.S.; Yaremchuk, A.D.; Protopopov, M.V.; Yarmoluk, S.M.; Tukalo, M.A. Identification of *Mycobacterium tuberculosis* Leucyl-TRNA Synthetase (LeuRS) Inhibitors among the Derivatives of 5-Phenylamino-2H-[1,2,4]Triazin-3-One. *J. Enzyme Inhib. Med. Chem.* **2016**, *31*, 201–207. [CrossRef] [PubMed]
12. Gudzera, O.I.; Golub, A.G.; Bdzhola, V.G.; Volynets, G.P.; Lukashov, S.S.; Kovalenko, O.P.; Kriklivyi, I.A.; Yaremchuk, A.D.; Starosyla, S.A.; Yarmoluk, S.M.; et al. Discovery of Potent Anti-Tuberculosis Agents Targeting Leucyl-TRNA Synthetase. *Bioorg. Med. Chem.* **2016**, *24*, 1023–1031. [CrossRef] [PubMed]
13. Mahapatra, M.K.; Karuppasamy, M. Fundamental Considerations in Drug Design Chapter Outline. In *Computer Aided Drug Design (CADD): From Ligand-Based Methods to Structure-Based Approaches*; Elsevier: Amsterdam, The Netherlands, 2022; pp. 17–55. [CrossRef]
14. Devi, B.; Vasishta, S.S.; Das, B.; Baidya, A.T.K.; Rampa, R.S.; Mahapatra, M.K.; Kumar, R. Integrated Use of Ligand and Structure-Based Virtual Screening, Molecular Dynamics, Free Energy Calculation and ADME Prediction for the Identification of Potential PTP1B Inhibitors. *Mol. Divers.* **2023**, *28*, 649–669. [CrossRef] [PubMed]
15. Mahapatra, M.K.; Bera, K.; Singh, D.V.; Kumar, R.; Kumar, M. In Silico Modelling and Molecular Dynamics Simulation Studies of Thiazolidine Based PTP1B Inhibitors. *J. Biomol. Struct. Dyn.* **2018**, *36*, 1195–1211. [CrossRef] [PubMed]
16. Mahapatra, M.K.; Kumar, R.; Kumar, M. N-Alkylated Thiazolidine-2,4-Dione Analogs as PTP1B Inhibitors: Synthesis, Biological Activity, and Docking Studies. *Med. Chem. Res.* **2017**, *26*, 1176–1183. [CrossRef]
17. Sterling, T.; Irwin, J.J. ZINC 15—Ligand Discovery for Everyone. *J. Chem. Inf. Model.* **2015**, *55*, 2324–2337. [CrossRef] [PubMed]
18. Schrodinger Release. 2021–2022 | Schrödinger [WWW Document]. 2021. Available online: https://www.schrodinger.com/releases/release-2021-2 (accessed on 20 October 2023).
19. Uniyal, A.; Mahapatra, M.K.; Tiwari, V.; Sandhir, R.; Kumar, R. Targeting SARS-CoV-2 Main Protease: Structure Based Virtual Screening, in Silico ADMET Studies and Molecular Dynamics Simulation for Identification of Potential Inhibitors. *J. Biomol. Struct. Dyn.* **2022**, *40*, 3609–3625. [CrossRef] [PubMed]
20. Friesner, R.A.; Banks, J.L.; Murphy, R.B.; Halgren, T.A.; Klicic, J.J.; Mainz, D.T.; Repasky, M.P.; Knoll, E.H.; Shelley, M.; Perry, J.K.; et al. Glide: A New Approach for Rapid, Accurate Docking and Scoring. 1. Method and Assessment of Docking Accuracy. *J. Med. Chem.* **2004**, *47*, 1739–1749. [CrossRef] [PubMed]
21. Friesner, R.A.; Murphy, R.B.; Repasky, M.P.; Frye, L.L.; Greenwood, J.R.; Halgren, T.A.; Sanschagrin, P.C.; Mainz, D.T. Extra Precision Glide: Docking and Scoring Incorporating a Model of Hydrophobic Enclosure for Protein-Ligand Complexes. *J. Med. Chem.* **2006**, *49*, 6177–6196. [CrossRef] [PubMed]
22. Mahapatra, M.K.; Kumar, R.; Kumar, M. Synthesis, Biological Evaluation and in Silico Studies of 5-(3-Methoxybenzylidene) Thiazolidine-2,4-Dione Analogues as PTP1B Inhibitors. *Bioorg. Chem.* **2017**, *71*, 1–9. [CrossRef] [PubMed]
23. QikProp | Schrödinger. Available online: https://www.schrodinger.com/products/qikprop (accessed on 20 October 2023).

24. Mahapatra, M.K.; Kumar, R.; Kumar, M. Exploring Sulfonate Esters of 5-Arylidene Thiazolidine-2,4-Diones as PTP1B Inhibitors with Anti-Hyperglycemic Activity. *Med. Chem. Res.* **2018**, *27*, 476–487. [CrossRef]
25. Bowers, K.J.; Chow, E.; Xu, H.; Dror, R.O.; Eastwood, M.P.; Gregersen, B.A.; Klepeis, J.L.; Kolossvary, I.; Moraes, M.A.; Sacerdoti, F.D.; et al. Scalable Algorithms for Molecular Dynamics Simulations on Commodity Clusters. In Proceedings of the 2006 ACM/IEEE Conference on Supercomputing, Tampa, FL, USA, 11–17 November 2006; p. 84-es. [CrossRef]
26. Rehman, Z.; Najmi, A. Exploring EGFR Inhibitors with the Aid of Virtual Screening, Docking, and Dynamics Simulation Studies. *J. Biomol. Struct. Dyn.* **2023**, 1–21. [CrossRef] [PubMed]
27. Imran, M.; Alotaibi, N.M.; Thabet, H.K.; Alruwaili, J.A.; Asdaq, S.M.B.; Eltaib, L.; Kamal, M.; Alshammari, A.B.H.; Alshammari, A.M.A.; Alshehri, A. Computer-Assisted Discovery of Safe and Effective DprE1/AaRSs Inhibitors against TB Utilizing Drug Repurposing Approach. *J. Infect. Public Health* **2023**, *16*, 554–572. [CrossRef] [PubMed]
28. Damian, L. Isothermal titration calorimetry for studying protein-ligand interactions. *Methods Mol. Biol.* **2013**, *1008*, 103–118. [CrossRef] [PubMed]
29. Chakraborty, P.; Bajeli, S.; Kaushal, D.; Radotra, B.D.; Kumar, A. Biofilm Formation in the Lung Contributes to Virulence and Drug Tolerance of Mycobacterium Tuberculosis. *Nat. Commun.* **2021**, *12*, 1606. [CrossRef] [PubMed]
30. Sambanthamoorthy, K.; Feng, X.; Patel, R.; Patel, S.; Paranavitana, C. Antimicrobial and antibiofilm potential of biosurfactants isolated from lactobacilli against multi-drug-resistant pathogens. *BMC Microbiol.* **2014**, *14*, 197. [CrossRef] [PubMed]
31. Sánchezm, J.G.; Kouznetsov, V.V. Antimycobacterial susceptibility testing methods for natural products research. *Braz. J. Microbiol.* **2010**, *41*, 270–277. [CrossRef] [PubMed]
32. Sambandan, D.; Dao, D.N.; Weinrick, B.C.; Vilchèze, C.; Gurcha, S.S.; Ojha, A.; Kremer, L.; Besra, G.S.; Hatfull, G.F.; Jacobs, W.R. Keto-Mycolic Acid-Dependent Pellicle Formation Confers Tolerance to Drug-Sensitive *Mycobacterium tuberculosis*. *MBio* **2013**, *4*, 10–128. [CrossRef]
33. Nayak, N. *Mycobacterium tuberculosis* Biofilm—A New Perspective. *Indian J. Tuberc.* **2015**, *62*, 4–6. [CrossRef] [PubMed]
34. Li, J.; Abel, R.; Zhu, K.; Cao, Y.; Zhao, S.; Friesner, R.A. The VSGB 2.0 model: A next generation energy model for high resolution protein structure modeling. *Proteins* **2011**, *79*, 2794–2812. [CrossRef] [PubMed]

biomolecules

MDPI

Article

Development of a Novel In Silico Classification Model to Assess Reactive Metabolite Formation in the Cysteine Trapping Assay and Investigation of Important Substructures

Yuki Umemori, Koichi Handa *, Saki Yoshimura, Michiharu Kageyama and Takeshi Iijima

DMPK Research Department, Teijin Institute for Bio-Medical Research, TEIJIN PHARMA LIMITED, 4-3-2 Asahigaoka, Hino-shi, Tokyo 191-8512, Japan; yuki.umemori@axcelead-twp.com (Y.U.); saki.yoshimura@axcelead-twp.com (S.Y.); michiharu.kageyama@axcelead-twp.com (M.K.); t.iijima@teijin.co.jp (T.I.)

* Correspondence: koichi.handa@axcelead-twp.com; Tel.: +81-42-586-8260

Abstract: Predicting whether a compound can cause drug-induced liver injury (DILI) is difficult due to the complexity of drug mechanism. The cysteine trapping assay is a method for detecting reactive metabolites that bind to microsomes covalently. However, it is cumbersome to use 35S isotope-labeled cysteine for this assay. Therefore, we constructed an in silico classification model for predicting a positive/negative outcome in the cysteine trapping assay. We collected 475 compounds (436 in-house compounds and 39 publicly available drugs) based on experimental data performed in this study, and the composition of the results showed 248 positives and 227 negatives. Using a Message Passing Neural Network (MPNN) and Random Forest (RF) with extended connectivity fingerprint (ECFP) 4, we built machine learning models to predict the covalent binding risk of compounds. In the time-split dataset, AUC-ROC of MPNN and RF were 0.625 and 0.559 in the hold-out test, restrictively. This result suggests that the MPNN model has a higher predictivity than RF in the time-split dataset. Hence, we conclude that the in silico MPNN classification model for the cysteine trapping assay has a better predictive power. Furthermore, most of the substructures that contributed positively to the cysteine trapping assay were consistent with previous results.

Keywords: cysteine trapping assay; drug-induced liver injury (DILI); idiosyncratic DILI; hepatotoxicity; message passing neural network; random forest; substructures; QSAR; toxicity; reactive metabolite

1. Introduction

The success rate of new drug development is low, and one of the major causes of failure in clinical trials is toxicity [1]. Hepatotoxicity or cardiotoxicity accounts for most of these failures [2]. Since clinical drug development is very expensive, it is important for pharmaceutical companies to predict the toxicity of candidate compounds in the early stages of drug discovery to exclude compounds with risks.

Hepatotoxicity, namely drug-induced liver injury (DILI), in clinical trials includes intrinsic DILI and idiosyncratic DILI. Intrinsic DILI is dose-dependent and can often be avoided with dose adjustment [3]. In contrast, idiosyncratic DILI is dose-independent and depends on the patient's genetic background, so its incidence is rare [4–6], and thus it is extremely difficult to predict its occurrence. In some instances, idiosyncratic DILI is discovered after a product is already on the market, leading to drug withdrawal [7]. Recently, there was a case of fasiglifam-induced liver damage in the late clinical stage, leading to discontinuation of the clinical trial [8].

Based on the mechanism, idiosyncratic DILI is classified into hepatocellular, cholestasis, and mixed types. Although the detailed molecular mechanisms of these are still largely unknown, in hepatocellular injury, reactive metabolites covalently bind to proteins, which are then processed by antigen-presenting cells, leading to abnormal compound-specific

T-cells. The derived "hapten concept" is well known [9]. In the cholestatic form, drugs can inhibit the bile salt export pump (BSEP), leading to the accumulation of bile acids in hepatocytes, resulting in toxicity [10,11]. In addition, the generation of reactive oxygen species (ROS) and mitochondrial disorders are intricately intertwined [12]. By screening compounds in the early stages of drug discovery based on these mechanisms that have been recently clarified, we can avoid fatal outcomes in the late stages of marketing and clinical trials.

To experimentally evaluate idiosyncratic DILI during the nonclinical stage, a method was developed in the 2000s to measure the amount of covalent binding to microsomal or hepatocyte proteins in vitro using radiolabeled compounds. This method uses the amount of covalent binding as a quantitative index of reactive metabolites, and also examines the relationship with the clinical dose [13–15]. In addition, it was also shown that quantification was improved by incorporating multifaceted indicators, such as transporter inhibitory activity of proteins like BSEP and mitochondrial damage [16]. An assay for quantitatively estimating the covalent bonding ability of this compound to a protein requires the use of a radiolabeled compound, which is difficult to handle in the initial stages of drug discovery due to costs incurred and convenience. Glutathione (GSH) systems using fluorescence have also been proposed [17], but at present, tests using radiolabeled substrates (such as [^{35}S] cysteine or [^{14}C] cyanide) are conducted due to their high sensitivity [18,19]. By using such experimental systems, it is possible to evaluate idiosyncratic DILI in the early stages of drug discovery, although it is limited to the hepatocellular disorder type. For the cholestasis type, an experimental system using a sandwich culture method that forms bile ducts in vitro has been established [20], although it has not yet attained a predictive power suitable for screening in the early stages of drug discovery due to the difficulty of the experiment and costs. Hence, to improve our ability to forecast the possibility of various forms of hepatotoxicity at an early stage, prediction by computational methods would be ideal.

A consortium of pharmaceutical companies is currently developing DILI-Sim®, which is a computational prediction method for DILI [21]. However, although DILI-Sim is excellent for prediction in the late clinical stage and for already-marketed products [overdosed acetaminophen [22], ubrogepant comparing with telecagepant and MK-3207 [23], a sublingual formulation of riluzole [24], tolvaptan and its metabolite [25], the amount of input information required is too large to apply it at the drug discovery stage. In contrast, in silico models using various machine learning algorithms have been developed for DILI prediction in the early stages of drug discovery. Until now, the development of prediction methods based on classical methods using the structure of compounds has been vigorously pursued [26–29]; however, there are also many multimodal methods being developed that combine various explanatory variables such as calculated values (target protein prediction) [30] or experimental values (gene expression, protein expression, imaging data, etc.) [29,31,32]. It should be noted that the dataset used (called LTKB) was labeled with clinical information (casualty, incidence, severity from trials, literature surveys, and reports [33]. Considering the difficulty of predicting DILI in clinics [34], intrinsic DILI and idiosyncratic DILI might overlap, making the task of predicting DILI much more difficult. Furthermore, recently, various machine learning approaches have been proposed for searching for optimal models by combining multiple algorithms and multiple descriptors [35], converting complex toxicity information into integrated descriptors [36], and precise prediction by multi-binary classification models [37].

In this study, we decided not to create a direct prediction model for DILI, considering that the mechanism of DILI is diverse. Instead, we focused on predicting the formation of reactive metabolites that can predict idiosyncratic DILI with high probability. It is possible to find a machine learning model for predicting the site of GSH conjugates [38]. Here, to achieve quantitative prediction, specifically, using the ^{35}S-cysteine trapping data acquired in-house, we created an in silico binary classification model using machine learning. Although it is not possible to directly predict DILI, this model can predict the presence or absence of reactive metabolites, which are a crucial predictor of idiosyncratic DILI. In addition, to make a more practical contribution to drug discovery, we also estimated the substructures that cause the generation of reactive metabolites.

2. Materials and Methods

2.1. Experimental Method

2.1.1. Sample Preparation

The reagents for making the reaction solution were phosphate buffer (pH 7.4; 100 mM), ethylenediaminetetraacetic acid (EDTA; 1 mM) (Tokyo Kasei, Tokyo, Japan), $MgCl_2$ hexahydrate (Fuji Film Wako Co., Ltd., Osaka, Japan, reagent special grade; 3 mM), glucose-6-phosphate (G6P; 5 mM) (Roche, Basel, Switzerland), G6P dehydrogenase (G6PDH; 1 IU) (Oriental Yeast Co., Ltd., Tokyo, Japan), human liver microsome solution (Xenotech, Kansas City, MO, USA; 2 mg/mL), and L-[35S]-Cysteine (PerkinElmer Japan Co., Ltd., Kanagawa, Japan; 3.7 MBq/mL), which were prepared using ultrapure water. The evaluation compound was dissolved in dimethyl sulfoxide (DMSO) and its concentration was adjusted to 4 mM. This was used as a substrate solution.

2.1.2. Reaction in Microsome

On ice, 239 µL of the reaction solution was dispensed into a glass tube and then 1.25 µL of the substrate solution was added, mixed, and placed in a water bath (PERSONAL-11, Taitec Co., Ltd., Saitama, Japan) at 37 °C. After pre-incubation for 5 min, 10 µL of 25 mM nicotinamide adenine dinucleotide phosphate (NADPH) (Roche, Basel, Switzerland) solution was added and mixed to a final concentration of 20 µM to initiate a metabolic reaction and incubated in a water bath at 37 °C for 2 h. After incubation, 500 µL of acetonitrile was added and mixed, and the tube was chilled on ice to deproteinize, after which 50 µL of 250 mM GSH solution was added and mixed to stop the reaction, and centrifuged (approximately 1500× *g*, 4°C, 10 min, CF16RXII, Hitachi, Ltd., Tokyo, Japan). After centrifugation, 750 µL of the supernatant was transferred to a new glass tube and concentrated by vacufugation (VC-96N, Taitec Co., Ltd., Saitama, Japan) for 2 h. After concentrating the solution, we added 100 µL of the redissolving solution (40% methanol solution), mixed it with a vortex (IS-MBI minimixer, Ikeda Rika Co., Ltd., Tokyo, Japan) to redissolve, and then the sample was placed in a filter tube [Merck Millipore, Darmstadt, Germany] and centrifuged (approximately 5000× *g*, room temperature, 10 min).

2.1.3. Measurement by HPLC with Radiomatic Detector

Both High Performance Liquid Chromatography (HPLC) (Nexera XR, Shimadzu Corporation, Kyoto, Japan) and a radiomatic detector (Radiomatic 610TR, PerkinElmer Japan Co., Ltd.) were used to analyze the prepared samples. The 35S content was detected by running the separated sample via the 45-min analytical method through a radiomatic detector. The mobile phase was composed of a gradient of 0.1% (*v*/*v*) formic acid/water and 0.1% (*v*/*v*) formic acid/acetonitrile, and the chromatographic separation was performed using a column (Synergi 4 µm Hydro-RP 80A 150 × 3.00 mm, Phenomenex, Torrance, CA, USA).

2.2. Compound Preparation

Cysteine trapping compound data (475 compounds) were used; 436 in-house compounds and 39 available marketed drugs were used. The composition of the experimental results showed 248 positives and 227 negatives, all of the result of (RI) integrated area is shown in Table S1. For in-house compounds, we used compound structures registered in the Teijin database. Structural information for known compounds was obtained as 2D SDF files using ChEMBL33 [39]. It can be seen from Figure S1 that the compounds used in this study mostly meet with the rule of five criteria for molecular weight (MW) and LogP [40]; and the values of polar surface area (PSA) are mostly similar to drugs which has good absorption into human body [41]. Our compounds are seemed to be drug-like and those are suitable for the dataset of this predictive model.

2.3. Compound Standardization through SMILES Representation

The above structural information was loaded into Maestro (Schrödinger suite 10.3) [42] and converted to canonical simplified molecular-input line-entry system (SMILES) without considering isomerization. These SMILES were used to calculate the molecular weight, polar surface area, and logP in Insight for excel [43]. These are shown as a frequency distribution graph in a Supplementary Materials (Figure S1).

2.4. Classification of Data by Principal Component Analysis (PCA) Plot

To confirm that the training and test compound structures used in the random and time-split analyses were sampled from the same compound space, PCA was performed using DataWarrior (Version 5.5.0) ([44]). In DataWarrior, the compound structure in canonical SMILES format was read as input, and then FragFP was calculated as a fingerprint. Using the values of this FragFP, PCA was performed with two components.

2.5. Model Building with Machine Learning

2.5.1. Overall Workflow

The construction of the machine learning model followed the steps depicted in Figure 1.

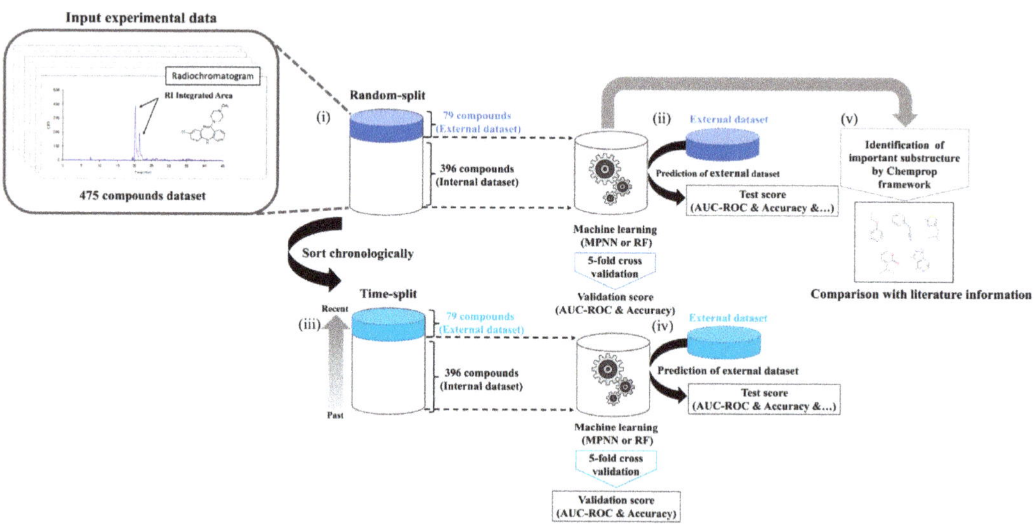

Figure 1. Workflow for building the cysteine trapping QSAR prediction model. (**i**) Structural information and RI (Radio Isotope) integrated area, which is counted as the radiochromatographic peaks of 475 compounds, including 436 in-house and 39 known compounds, were used for the dataset. Of the 475 compounds, 1/6 or 79 compounds were randomly extracted and used as an external dataset, and the remaining 396 compounds were used as a training dataset. (**ii**) Next, machine learning was performed on 396 compounds, 5-fold cross-validation (CV) was performed to calculate the validation score of the prediction model, and a hold-out test was performed on the external dataset to calculate the test score. (**iii**) The 475 compounds were rearranged in chronological order, and the most recent 79 compounds were extracted and used as an external dataset. (**iv**) Similar to (**iii**), machine learning was performed with 396 compounds on the time-split dataset, 5-fold CV was performed to calculate the validation score of the prediction model, hold-out test was performed on the external dataset, and the test score was calculated. (**v**) Using the dataset of 396 compounds from random splitting, we estimated the substructure used for positive determination.

2.5.2. Chemical Structure Fingerprints

For machine learning modeling, chemical fingerprintss and chemical structure as graphs were used. ECFP4 (2048 bit, radius: 2) [45] was calculated using the python RD-kit (2020.09.01) [46] in Chem function and AllChem, and for chemical graphs, canonical SMILES were used. Regarding chemical fingerprints, ECFP4 is a well-known fingerprint that can represent compounds better than the other fingerprints, especially for drug candidates [47]; therefore, considering its practical usage in the drug discovery process, we decided to use it.

2.5.3. Objective Variable (Cysteine Trapping Posi/Nega Classification)

It has been reported that there is a correlation between the total radio isotope (RI) integrated area (unit; count) in cysteine trapping and the amount of covalent binding in microsomes (unit; pmol/g) in in vitro covalent binding tests, and there is a risk of idiosyncratic DILI at 50 pmol/g or more [18]. Considering this, the amount of covalent binding in microsomes in the in vitro covalent binding test of 50 pmol/g or more was judged as a risk. The threshold of the total RI integrated area in cysteine trapping at that time was about 1000 counts. Therefore, we labeled compounds with more than 1000 counts of RI integrated area as positive, and those with less than 1000 counts as negative.

2.5.4. Machine Learning Models

We utilized the following models: Random Forests (RF) [48] using chemical fingerprints as input, and Message Passing Neural Network (MPNN) [49] using graphs as input. We used RF because it is a stable machine learning model that has already been used in various published articles [50–52]. MPNN was also used because excellent predictive MPNN models have been proposed in recent years [53–55]. The Python (version 3.7.10) scikit-learn (version 0.24.2) and the Python (version 3.7.10) Chemprop (version 1.3.1) library chemprop function were used for RF and MPNN, respectively. The parameters were set to their defaults [56].

2.5.5. Model Validation

To demonstrate the validity of the model, we divided the data set using two methods for hold-out tests (random selected and time-split) and performed 5-fold cross-validation (CV). First, in random-split, 5/6 of the total data set was randomly selected as a training dataset, and the remaining 1/6 was selected as a hold-out test. Next, in time-split, the in-house compounds were sorted in the order of synthesis date, and 5/6 of the total data from the earliest was selected as a training dataset, and the remaining 1/6 of total data from the latest was selected as a hold-out test (Figure 1i–iv).

2.5.6. Estimation of Important Substructures

We estimated the important substructures of the cysteine trapping assay using the MPNN model built in this study by the Monte Carlo Tree search under Chemprop framework (Figure 1v) [57]. Regarding the Monte Carlo Tree search, chemprop.interpret.py in the Python (version 3.7.10) Chemprop (version 1.3.1) was used, and the parameters were set to their defaults (Input: smiles, Number of message passing steps: 3, Batch size: 50. Number of epochs to run: 30, Learning rate: 0.001) we currently restricted the rationale to have maximum 20 atoms and minimum 8 atoms [56].

2.5.7. Metrics to Compare Each In Silico Model

In the 5-fold CV, AUC-ROC and accuracy were calculated to grasp the trend of model accuracy. In the hold-out test, in addition to AUC-ROC and accuracy, the following indices (1) to (8) were calculated from the confusion matrix to examine the extrapolation of the model in detail.

TN: True Negative, FP: False Positive, FN: False Negative, TP: True Positive, MCC: Matthews Correlation Coefficient

$$\text{Accuracy} = \frac{\text{TP} + \text{TN}}{\text{TP} + \text{FP} + \text{TN} + \text{FN}} \tag{1}$$

$$\text{Precision} = \frac{\text{TP}}{\text{TP} + \text{FP}} \tag{2}$$

$$\text{Sensitivity} = \frac{\text{TP}}{\text{TP} + \text{FN}} \tag{3}$$

$$\text{Specificity} = \frac{\text{TN}}{\text{TN} + \text{FP}} \tag{4}$$

$$\text{Youden's index} = \text{Sensitivity} + \text{Specificity} - 1 \tag{5}$$

$$\text{MCC} = \frac{\text{TP} \times \text{TN} - \text{FP} \times \text{FN}}{\sqrt{(\text{TP} + \text{FP})(\text{TP} + \text{FN})(\text{TN} + \text{FP})(\text{TN} + \text{FN})}} \tag{6}$$

$$\text{False rate} = \frac{\text{FN} + \text{FP}}{(\text{TP} + \text{FP} + \text{TN} + \text{FN})} \tag{7}$$

$$\text{F} - \text{measure} = \frac{2 \times \text{Precision} \times \text{Sensitivity}}{\text{Precision} + \text{Sensitivity}} \tag{8}$$

3. Results and Discussion

3.1. PCA Distribution of Training and Test Dataset

First, a random selection of 79 compounds was performed from the total pool of 475 compounds to assess the distribution of training and test dataset. These 79 compounds were used as an external dataset in a random-split. Additionally, the most recent 79 compounds were extracted for an external dataset as a time-split approach. PCA using FragFP was performed for the random-split and time-split to investigate the chemical space of the training and test dataset, respectively. In the random-split, the compound distribution of the training and test sets matched and was evenly distributed (Figure 2a). A bias was observed in the distribution of the test set in the time-split, likely due to recently synthesized compounds having similar targets and compound structures (Figure 2b). These findings indicate evident differences in the data selection processes between the random-split and time-split approaches, highlighting the importance of model evaluation in the time-split [58]. Furthermore, it was confirmed that the appropriate test/train selection was performed to construct a model in which the compound space remained consistent, even in the time-split, as some training compounds matched the compound distribution of the test dataset. However, it should be noted that all the compounds in the dataset are pharmaceuticals. Additionally, PCA analysis is also performed using ECFP4 as descriptor and is shown in Figure S2. (Although the conclusion was the same of FragFP, we do not discuss it due to its low contribution ratio in PC1 and PC2.)

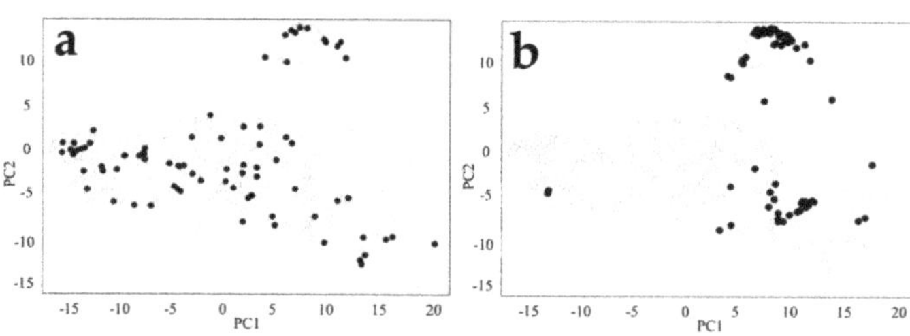

Figure 2. PCA plot of training and test dataset (**a**) Random-split (**b**) Time-split. White and black circles indicate compounds used in the training and external datasets, respectively. The contribution ratio of PC1 and PC2 is 26%.

3.2. Evaluation with Random-Split

To evaluate the models under the random-split, 5-fold CV was performed on both the MPNN and RF models. Table 1 shows that the ROC-AUC and accuracy of the MPNN model were 0.698 ± 0.08 and 0.668 ± 0.08, whereas those of the RF model were 0.811 ± 0.03 and 0.752 ± 0.02, respectively. Table 2 and Figure 3a,b show the prediction results for the 79 randomly extracted compounds that were included in the external dataset for each model. The ROC-AUC and accuracy of the MPNN model were 0.742 and 0.696, whereas those of the RF model were 0.819 and 0.734, respectively. Notably, the precision of the RF model was 0.789, whereas that of the MPNN model was 0.732. Both models had the same sensitivity value of 0.698. With both models surpassing an ROC-AUC of 0.70 and exhibiting favorable metrics, including sensitivity, the MPNN and RF models showed the ability to effectively predict the risk of cysteine trapping using only compound structural information. This indicates that time and cost can be reduced because experimental data are not required [49].

Table 1. Metrics calculation for 5-fold cross validation in Message Passing Neural Network (MPNN) and Random Forest (RF) models.

Metrics	**Metrics**		**Models**				
			MPNN			**RF**	
Random-split	ROC-AUC	0.698	±	0.08	0.811	±	0.03
	Accuracy	0.668	±	0.08	0.752	±	0.02
Time-Split	ROC-AUC	0.729	±	0.05	0.617	±	0.17
	Accuracy	0.668	±	0.10	0.556	±	0.09
					Mean value	±	Standard deviation

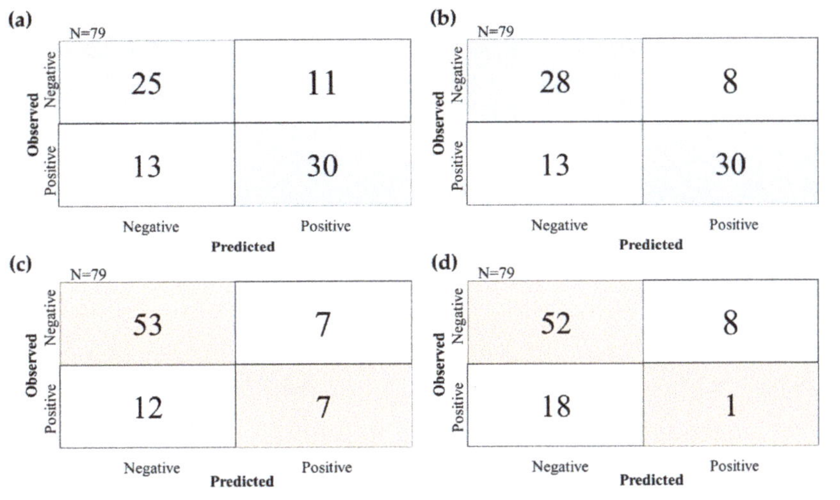

Figure 3. Confusion matrix of prediction with external dataset in random-split and time-split. (**a**) Prediction result of hold-out test by MPNN model at random-split, (**b**) Prediction result of hold-out test by RF model at random-split, (**c**) Prediction result of hold-out test by MPNN model at time-split, (**d**) Prediction result of hold-out test by RF model at time-split.

Table 2. Metrics calculation for predicting external dataset of Message Passing Neural Network (MPNN) and Random Forest (RF) Models.

Dataset	Models	ROC-AUC	Accuracy	Precision	Sensitivity	Specificity	Youden's Index	MCC	False Rate	F-Measure
Random-split	MPNN	0.742	0.696	0.732	0.698	0.694	0.392	0.391	0.304	0.714
	RF	0.819	0.734	0.789	0.698	0.778	0.475	0.474	0.266	0.741
Time-split	MPNN	0.625	0.759	0.500	0.368	0.883	0.252	0.282	0.241	0.424
	RF	0.559	0.671	0.111	0.053	0.867	−0.081	−0.109	0.329	0.071

3.3. Evaluation with Time-Split

The time-split approach was used in this study because it is a more legitimate method for evaluating models than the random-split approach, which has been used in many studies recently [59–61]. To validate the predictive performance considering the practical aspects, a dataset of 79 recently synthesized compounds was selected as an external dataset, whereas the remaining 396 compounds were used for a 5-fold CV. The MPNN and RF models were investigated. Table 1 shows that the ROC-AUC and accuracy of the MPNN model were 0.729 ± 0.05 and 0.668 ± 0.10, while those of the RF model were 0.617 ± 0.17 and 0.556 ± 0.09 respectively, illustrating the superior accuracy of the MPNN model. The prediction results for the external dataset using the 79 chronologically extracted compounds are shown in Table 2 and Figure 3c,d. The ROC-AUC and accuracy of the MPNN model were 0.625 and 0.759, whereas those of the RF model were 0.559 and 0.671, respectively. These results indicate the higher accuracy of the MPNN model, particularly when the objective variable is imbalanced. Furthermore, the RF model exhibited a low sensitivity of 0.053, whereas the MPNN model demonstrated a higher value of 0.368, indicating that the MPNN model had an improved prediction accuracy for the compounds at a risk of generating reactive metabolites. Moreover, when evaluating the time-split dataset, the F-measure values were 0.424 and 0.071 for the MPNN and RF models, respectively, which were lower than those from the evaluation of the random-split dataset (0.714 and 0.741). However, as the F-measure is influenced by the imbalance between positive and negative objective variables in the dataset, it is assumed that this difference is due to bias of the time-split dataset (Table 3). Conversely, the MCC scores, which are independent of dataset bias, were 0.282 and −0.109 for the MPNN and RF models, respectively. The higher MCC value for the MPNN model, close to the score 0.391, indicates that the graph information captured by the MPNN model effectively extracts compound structure features without relying on dataset bias, in contrast to ECFP4. These findings align with previous research supporting the advantages of the MPNN model [53,54].

Table 3. Composition in each positive and negative dataset.

Dataset		Positive		Negative	
Random	Train	205	(52%)	191	(48%)
	Test	43	(54%)	36	(46%)
Time	Train	229	(58%)	167	(42%)
	Test	19	(24%)	60	(76%)

3.4. Estimation of Important Substructures

We determined that the MPNN model exhibited superior prediction accuracy and used it to estimate the substructures contributing to positive classification. Among the 396 compounds in the random-split training dataset, 33 market compounds were included, of which 12 yielded positive results. The MPNN model classified 6 compounds as positive correctly, of which 5 were associated with substructures reported to be involved in reactive metabolite formation. The estimated substructures of 3 compounds aligned with the literature references, whereas 2 compounds had novel substructures. In Figure 4, the structures of market drugs and putative substructures in MPNN of these five compounds are shown, along with the putative structures contributing to positive classification and the substructures reported in the literature for reactive metabolite formation.

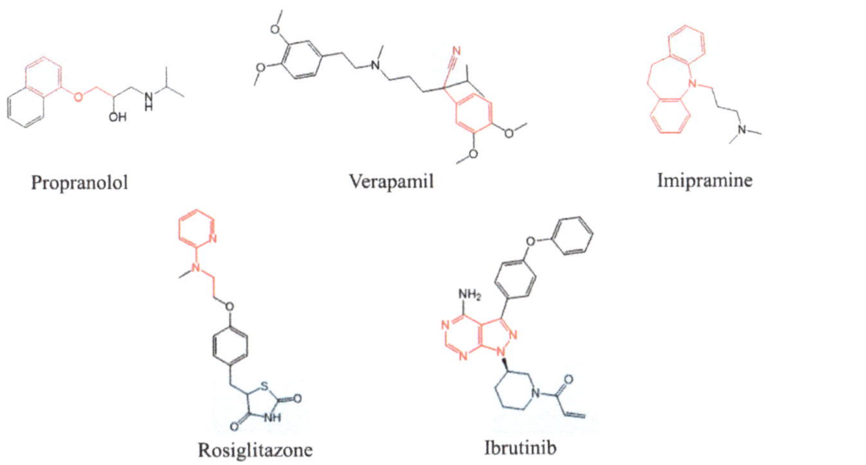

Figure 4. Structures of market drugs and putative substructures in MPNN. The blue highlighted part indicates the substructure reported in the literature to be involved in the generation of reactive metabolites, and the red sticks indicate the substructure predicted to have a positive contribution in the MPNN model.

3.4.1. Propranolol

Propranolol is metabolized by CYP2D in rat liver microsomes and leads to the formation of 4-hydroxypropranolol (4-OH-PL), which is further metabolized to 1,4-naphthoquinone (1,4-NQ) [62]. The ethoxybenzene and 4-OH-PL structures inferred in this study overlapped with the structures reported in the literature.

3.4.2. Verapamil

A putative substructure containing a cyano group, 2-phenylacetonitrile, was proposed for verapamil. Cyano groups generate adducts non-enzymatically, independent of the attached structure [63]. Thus, the cyano group in the estimated substructure aligns with the reported substructure.

3.4.3. Imipramine

Imipramine is metabolized by CYP2D in rat liver microsomes to form 2-hydroxy imipramine, which undergoes an arene-oxide intermediate formation, and covalently binds to the CYP2D protein as a reactive metabolite [64,65]. The MPNN model classifies imipramine as positive, with a putative substructure of 5-methyl-10,11-dihydro-5H-dibenzo[b,f]azepine. This substructure inferred in this study overlap with the structure reported in the literature.

3.4.4. Rosiglitazone

Rosiglitazone belongs to the thiazolidinedione class of antidiabetic drugs, along with troglitazone and pioglitazone. Troglitazone, the first therapeutic drug for insulin-resistant diabetes, was discontinued due to an idiosyncratic drug toxicity (IDT) associated with reactive metabolites. Several substructures of troglitazone, such as isocyanate-type metabolites, trigger IDT [66]. Initially, we hypothesized that the thiazolidinedione ring, which is a common structure among thiazolidinedione antidiabetic drugs, contributed to the reactive metabolites of rosiglitazone [67–69]. However, the MPNN model deduced a unique structure of N-ethylpyridin-2-amine, which differs from the thiazolidinedione ring. Although these three compounds exhibit covalent binding amounts of pmol/g in covalent binding assay reports [14], troglitazone and pioglitazone tested negative in the cysteine trapping assay due to the threshold area being less than 1000. In addition to the previously reported thiazolidione ring, the newly estimated N-ethylpyridin-2-amine may also contribute to the production of reactive metabolites.

3.4.5. Ibrutinib

Ibrutinib, a first-in-class inhibitor of Bruton's tyrosine kinase (BTK), irreversibly binds to cysteine 481 in the active site of BTK. 1-(Piperidin-1-yl) prop-2-en-1-one is the covalent binding site for BTK. However, the deduced structure in this study, the pyrazolopyrimidine ring, is in a different position. This difference in estimated structure arises from the fact that this model only predicts cysteine trapping scores and estimates the contributing structures [70]. As the system involves covalent bonding to BTK and differs from the experimental system of the cysteine trapping assay in this study, the pyrazolopyrimidine ring suggested here may contribute to the formation of new reactive metabolites.

4. Conclusions

The success rate of new drug development remains low, and DILI is often cited as a significant contributor to clinical trial failures, particularly the challenging task of predicting idiosyncratic DILI risks in the early stages of drug discovery. In this study, we focused on the cysteine trapping assay as a method for detecting reactive metabolites and developed a in silico binary classification model using in-house cysteine trapping data. To evaluate the model's accuracy, we employed two different dataset splitting approaches: a random-split for prediction accuracy using the MPNN prediction model and the RF model, and a more practical time-split. The MPNN model exhibited strong prediction accuracy in both the random-split and time-split scenarios, indicating its ability to extract graph-based structural information even when faced with dataset biases commonly encountered in real-world drug discovery settings. Furthermore, concerning the identified substructures used for the classification of known compounds, we discovered consistent findings with substructures reported in the literature known to carry a risk of generating reactive metabolites. However, the size of dataset used in this study is relatively small for machine learning in drug discovery, raising concerns about the model's generalizability and robustness. Expanding the dataset with more diverse chemical structures could improve the predictive performance and reliability. Nevertheless, this can contribute to the drug discovery process in a fit-for-purpose situations; for example, in the very early stage, thousands of compounds as the idea of medicinal chemists should be evaluated by this model instead of real experiments. Furthermore, this novel in silico model enables the extraction of the covalent binding risk, which is one of the factors for idiosyncratic DILI solely derived from structural information, offering potential applications in the early stages of drug discovery without the need for experimental data acquisition.

Supplementary Materials: The following supporting information can be downloaded at: https://www.mdpi.com/article/10.3390/biom14050535/s1, Table S1. Dataset for cysteine trapping assay (475 compounds), Figure S1. Distribution of physicochemical parameters, Figure S2. PCA plot of training and test dataset using Extended Connectivity FingerPrint 4 (a: Random-split b: Time-split).

Author Contributions: Y.U. and K.H., participated in research design, Y.U. and S.Y., wrote the manuscript, Y.U. and S.Y., collected data, K.H. and M.K., revised the manuscript. T.I. supervised the project. All authors have read and agreed to the published version of the manuscript.

Funding: The authors received no financial support for the research, authorship, and/or publication of this article.

Data Availability Statement: We can share publicly available compounds' dataset.

Acknowledgments: We thank Ryo Matsuyama and Akiko Watanabe who are the members of Toxicology & DMPK Research Department, Teijin Institute for Bio-medical Research, TEIJIN PHARMA LIMITED.

Conflicts of Interest: All authors are employed by TEIJIN Pharma Ltd.

References

1. Sun, D.; Gao, W.; Hu, H.; Zhou, S. Why 90% of clinical drug development fails and how to improve it? *Acta Pharm. Sin. B* **2022**, *12*, 3049–3062. [CrossRef] [PubMed]
2. Hornberg, J.J.; Laursen, M.; Brenden, N.; Persson, M.; Thougaard, A.V.; Toft, D.B.; Mow, T. Exploratory toxicology as an integrated part of drug discovery. Part I: Why and how. *Drug Discov. Today* **2014**, *19*, 1131–1136. [CrossRef] [PubMed]
3. Mosedale, M.; Watkins, P.B. Drug-induced liver injury: Advances in mechanistic understanding that will inform risk management. *Clin. Pharmacol. Ther.* **2017**, *101*, 469–480. [CrossRef] [PubMed]
4. De Abajo, F.J.; Montero, D.; Madurga, M.; García Rodríguez, L.A. Acute and clinically relevant drug-induced liver injury: A population based case-control study. *Br. J. Clin. Pharmacol.* **2004**, *58*, 71–80. [CrossRef] [PubMed]
5. Shen, T.; Liu, Y.; Shang, J.; Xie, Q.; Li, J.; Yan, M.; Xu, J.; Niu, J.; Liu, J.; Watkins, P.B.; et al. Incidence and etiology of drug-induced liver injury in Mainland China. *Gastroenterology* **2019**, *156*, 2230–2241.e11. [CrossRef] [PubMed]
6. Lee, W.M. Drug-induced acute liver failure. *Clin. Liver Dis.* **2013**, *17*, 575–586. [CrossRef] [PubMed]
7. Greene, P.J.; La Vaque, T.J. Postmarketing surveillance and black box warnings. *JAMA* **2002**, *288*, 957–958, author reply 958. [PubMed]
8. Marcinak, J.F.; Munsaka, M.S.; Watkins, P.B.; Ohira, T.; Smith, N. Liver safety of Fasiglifam (TAK-875) in patients with Type 2 diabetes: Review of the global clinical trial experience. *Drug Saf.* **2018**, *41*, 625–640, Erratum in *Drug Saf.* **2018**, *41*, 1431–1437. [CrossRef] [PubMed]
9. Johnston, A.; Uetrecht, J. Current understanding of the mechanisms of idiosyncratic drug-induced agranulocytosis. *Expert Opin. Drug Metab. Toxicol.* **2015**, *11*, 243–257. [CrossRef] [PubMed]
10. Dawson, S.; Stahl, S.; Paul, N.; Barber, J.; Kenna, J.G. In vitro inhibition of the bile salt export pump correlates with risk of cholestatic drug-induced liver injury in humans. *Drug Metab. Dispos.* **2012**, *40*, 130–138. [CrossRef] [PubMed]
11. Morgan, R.E.; van Staden, C.J.; Chen, Y.; Kalyanaraman, N.; Kalanzi, J.; Dunn, R.T., 2nd; Afshari, C.A.; Hamadeh, H.K. A multifactorial approach to hepatobiliary transporter assessment enables improved therapeutic compound development. *Toxicol. Sci.* **2013**, *136*, 216–241. [CrossRef] [PubMed]
12. Ye, H.; Nelson, L.J.; Gómez Del Moral, M.; Martínez-Naves, E.; Cubero, F.J. Dissecting the molecular pathophysiology of drug-induced liver injury. *World J. Gastroenterol.* **2018**, *24*, 1373–1385. [CrossRef] [PubMed]
13. Obach, R.S.; Kalgutkar, A.S.; Soglia, J.R.; Zhao, S.X. Can in vitro metabolism-dependent covalent binding data in liver microsomes distinguish hepatotoxic from nonhepatotoxic drugs? An analysis of 18 drugs with consideration of intrinsic clearance and daily dose. *Chem. Res. Toxicol.* **2008**, *21*, 1814–1822. [CrossRef] [PubMed]
14. Usui, T.; Mise, M.; Hashizume, T.; Yabuki, M.; Komuro, S. Evaluation of the potential for drug-induced liver injury based on in vitro covalent binding to human liver proteins. *Drug Metab. Dispos.* **2009**, *37*, 2383–2392. [CrossRef] [PubMed]
15. Nakayama, S.; Atsumi, R.; Takakusa, H.; Kobayashi, Y.; Kurihara, A.; Nagai, Y.; Nakai, D.; Okazaki, O. A zone classification system for risk assessment of idiosyncratic drug toxicity using daily dose and covalent binding. *Drug Metab. Dispos.* **2009**, *37*, 1970–1977. [CrossRef] [PubMed]
16. Thompson, R.A.; Isin, E.M.; Li, Y.; Weidolf, L.; Page, K.; Wilson, I.; Swallow, S.; Middleton, B.; Stahl, S.; Foster, A.J.; et al. In vitro approach to assess the potential for risk of idiosyncratic adverse reactions caused by candidate drugs. *Chem. Res. Toxicol.* **2012**, *25*, 1616–1632. [CrossRef] [PubMed]
17. Gan, J.; Harper, T.W.; Hsueh, M.M.; Qu, Q.; Humphreys, W.G. Dansyl glutathione as a trapping agent for the quantitative estimation and identification of reactive metabolites. *Chem. Res. Toxicol.* **2005**, *18*, 896–903. [CrossRef] [PubMed]
18. Inoue, K.; Shibata, Y.; Takahashi, H.; Ohe, T.; Chiba, M.; Ishii, Y. A trapping method for semi-quantitative assessment of reactive metabolite formation using [^{35}S]cysteine and [^{14}C]cyanide. *Drug Metab. Pharmacokinet.* **2009**, *24*, 245–254. [CrossRef] [PubMed]
19. Kakutani, N.; Nanayama, T.; Nomura, Y. Novel risk assessment of reactive metabolites from discovery to clinical stage. *J. Toxicol. Sci.* **2019**, *44*, 201–211. [CrossRef] [PubMed]
20. Ogimura, E.; Sekine, S.; Horie, T. Bile salt export pump inhibitors are associated with bile acid-dependent drug-induced toxicity in sandwich-cultured hepatocytes. *Biochem. Biophys. Res. Commun.* **2011**, *416*, 313–317. [CrossRef] [PubMed]
21. DILIsym®. Available online: https://www.simulations-plus.com/software/dilisym/ (accessed on 12 July 2023).
22. Woodhead, J.L.; Howell, B.A.; Yang, Y.; Harrill, A.H.; Clewell, H.J., 3rd; Andersen, M.E.; Siler, S.Q.; Watkins, P.B. An analysis of N-acetylCysteine treatment for acetaminophen overdose using a systems model of drug-induced liver injury. *J. Pharmacol. Exp. Ther.* **2012**, *342*, 529–540. [CrossRef] [PubMed]
23. Smith, B.; Rowe, J.; Watkins, P.B.; Ashina, M.; Woodhead, J.L.; Sistare, F.D.; Goadsby, P.J. Mechanistic investigations support liver safety of ubrogepant. *Toxicol. Sci.* **2020**, *177*, 84–93. [CrossRef] [PubMed]
24. Longo, D.M.; Shoda, L.K.M.; Howell, B.A.; Coric, V.; Berman, R.M.; Qureshi, I.A. Assessing Effects of BHV-0223 40 mg Zydis sublingual Formulation and riluzole 50 mg Oral Tablet on liver function test Parameters Utilizing DILIsym. *Toxicol. Sci.* **2020**, *175*, 292–300. [CrossRef] [PubMed]
25. Beaudoin, J.J.; Brock, W.J.; Watkins, P.B.; Brouwer, K.L.R. Quantitative systems toxicology modeling predicts that reduced biliary efflux contributes to tolvaptan hepatotoxicity. *Clin. Pharmacol. Ther.* **2021**, *109*, 433–442. [CrossRef] [PubMed]

26. Chen, M.; Hong, H.; Fang, H.; Kelly, R.; Zhou, G.; Borlak, J.; Tong, W. Quantitative structure-activity relationship models for predicting drug-induced liver injury based on FDA-approved drug labeling annotation and using a large collection of drugs. *Toxicol. Sci.* **2013**, *136*, 242–249. [CrossRef] [PubMed]
27. Xu, Y.; Dai, Z.; Chen, F.; Gao, S.; Pei, J.; Lai, L. Deep learning for drug-induced liver injury. *J. Chem. Inf. Model.* **2015**, *55*, 2085–2093. [CrossRef] [PubMed]
28. Nguyen-Vo, T.H.; Nguyen, L.; Do, N.; Le, P.H.; Nguyen, T.N.; Nguyen, B.P.; Le, L. Predicting drug-induced liver injury using convolutional neural network and molecular fingerprint-embedded features. *ACS Omega* **2020**, *5*, 25432–25439. [CrossRef] [PubMed]
29. Li, T.; Tong, W.; Roberts, R.; Liu, Z.; Thakkar, S. Deep learning on high-throughput transcriptomics to predict drug-induced liver injury. *Front. Bioeng. Biotechnol.* **2020**, *8*, 562677. [CrossRef] [PubMed]
30. Liu, A.; Walter, M.; Wright, P.; Bartosik, A.; Dolciami, D.; Elbasir, A.; Yang, H.; Bender, A. Prediction and mechanistic analysis of drug-induced liver injury (DILI) based on chemical structure. *Biol. Direct* **2021**, *16*, 6. [CrossRef] [PubMed]
31. Liu, X.; Zheng, D.; Zhong, Y.; Xia, Z.; Luo, H.; Weng, Z. Machine-learning prediction of oral drug-induced liver injury (DILI) via multiple features and endpoints. *BioMed Res. Int.* **2020**, *2020*, 4795140. [CrossRef] [PubMed]
32. Zhu, X.W.; Sedykh, A.; Liu, S.S. Hybrid in silico models for drug-induced liver injury using chemical descriptors and in vitro cell-imaging information. *J. Appl. Toxicol.* **2014**, *34*, 281–288. [CrossRef] [PubMed]
33. LTKB Benchmark Dataset. 2023. Available online: https://www.fda.gov/science-research/liver-toxicity-knowledge-base-ltkb/ltkb-benchmark-dataset. (accessed on 12 July 2023).
34. Katarey, D.; Verma, S. Drug-induced liver injury. *Clin. Med.* **2016**, *16* (Suppl. 6), s104–s109. [CrossRef] [PubMed]
35. Yang, Q.; Zhang, S.; Li, Y. Deep learning algorithm based on molecular fingerprint for prediction of drug-induced liver injury. *Toxicology* **2024**, *502*, 153736. [CrossRef] [PubMed]
36. Gu, Y.; Wang, Y.; Wu, Z.; Li, W.; Liu, G.; Tang, Y. In silico prediction of drug-induced liver injury with a complementary integration strategy based on hybrid representation. *Mol. Inform.* **2023**, *42*, e2200284. [CrossRef] [PubMed]
37. Kelleci Çeli, K.F.; Karaduman, G. Machine learning-based prediction of drug-induced hepatotoxicity: An OvA-QSTR approach. *J. Chem. Inf. Model.* **2023**, *63*, 4602–4614. [CrossRef] [PubMed]
38. Hughes, T.B.; Miller, G.P.; Swamidass, S.J. Site of reactivity models predict molecular reactivity of diverse chemicals with glutathione. *Chem. Res. Toxicol.* **2015**, *28*, 797–809. [CrossRef] [PubMed]
39. CHEMBL. Available online: https://www.ebi.ac.uk/chembl/ (accessed on 4 July 2023).
40. Lipinski, C.A. Lead- and drug-like compounds: The rule-of-five revolution. *Drug Discov. Today Technol.* **2004**, *1*, 337–341. [CrossRef] [PubMed]
41. Kelder, J.; Grootenhuis, P.D.; Bayada, D.M.; Delbressine, L.P.; Ploemen, J.P. Polar molecular surface as a dominating determinant for oral absorption and brain penetration of drugs. *Pharm. Res.* **1999**, *16*, 1514–1519. [CrossRef] [PubMed]
42. Maestro. Life Science: Maestro-Schrödinger. Available online: https://www.schrodinger.com/platform/products/maestro/ (accessed on 22 March 2024).
43. Insight for Excel. Available online: https://www.3ds.com/products/biovia/insight/insight-excel (accessed on 6 March 2024).
44. Datawarrior. Available online: www.openmolecules.org/ (accessed on 4 October 2023).
45. Rogers, D.; Hahn, M. Extended-connectivity fingerprints. *J. Chem. Inf. Model.* **2010**, *50*, 742–754. [CrossRef] [PubMed]
46. RD-kit. Available online: https://www.rdkit.org/docs/index.html# (accessed on 25 March 2022).
47. O'Boyle, N.M.; Sayle, R.A. Comparing structural fingerprints using a literature-based similarity benchmark. *J. Cheminform.* **2016**, *8*, 36. [CrossRef] [PubMed]
48. Breiman, L. Random forests. *Mach. Learn.* **2001**, *45*, 5–32. [CrossRef]
49. Yang, K.; Swanson, K.; Jin, W.; Coley, C.; Eiden, P.; Gao, H.; Guzman-Perez, A.; Hopper, T.; Kelley, B.; Mathea, M.; et al. Analyzing learned molecular representations for property prediction. *J. Chem. Inf. Model.* **2019**, *59*, 3370–3388. [CrossRef] [PubMed]
50. Xu, Y.; Ma, J.; Liaw, A.; Sheridan, R.P.; Svetnik, V. Demystifying multitask deep neural networks for quantitative structure-activity relationships. *J. Chem. Inf. Model.* **2017**, *57*, 2490–2504. [CrossRef] [PubMed]
51. Handa, K.; Sugiyama, S.; Kageyama, M.; Iijima, T. Combined data-driven and mechanism-based approaches for human-intestinal-absorption prediction in the early drug-discovery stage. *Digit. Discov.* **2023**, *2*, 1577–1588. [CrossRef]
52. Göller, A.H.; Kuhnke, L.; Montanari, F.; Bonin, A.; Schneckener, S.; Ter Laak, A.; Wichard, J.; Lobell, M.; Hillisch, A. Bayer's in silico ADMET platform: A journey of machine learning over the past two decades. *Drug Discov. Today* **2020**, *25*, 1702–1709. [CrossRef]
53. Jiang, D.; Wu, Z.; Hsieh, C.Y.; Chen, G.; Liao, B.; Wang, Z.; Shen, C.; Cao, D.; Wu, J.; Hou, T. Could graph neural networks learn better molecular representation for drug discovery? A comparison study of descriptor-based and graph-based models. *J. Cheminform.* **2021**, *13*, 12. [CrossRef] [PubMed]
54. Isert, C.; Kromann, J.C.; Stiefl, N.; Schneider, G.; Lewis, R.A. Machine learning for fast, quantum mechanics-based approximation of drug lipophilicity. *ACS Omega* **2023**, *8*, 2046–2056. [CrossRef] [PubMed]
55. Heid, E.; Greenman, K.P.; Chung, Y.; Li, S.C.; Graff, D.E.; Vermeire, F.H.; Wu, H.; Green, W.H.; McGill, C.J. Chemprop: A Machine Learning Package for Chemical Property Prediction. *J. Chem. Inf. Model.* **2023**, *64*, 9–17. [CrossRef] [PubMed]
56. Chemprop. Available online: https://chemprop.readthedocs.io/en/latest/# (accessed on 25 March 2023).

57. Jin, W.; Barzilay, R.; Jaakkola, T. Multi-objective molecule generation using interpretable substructures. *arXiv* **2020**, arXiv:2002.03244v3.
58. Sheridan, R.P. Time-split cross-validation as a method for estimating the goodness of prospective prediction. *J. Chem. Inf. Model.* **2013**, *53*, 783–790. [CrossRef] [PubMed]
59. Fang, C.; Wang, Y.; Grater, R.; Kapadnis, S.; Black, C.; Trapa, P.; Sciabola, S. Prospective validation of machine learning algorithms for absorption, distribution, metabolism, and excretion prediction: An industrial perspective. *J. Chem. Inf. Model.* **2023**, *63*, 3263–3274. [CrossRef]
60. Di Lascio, E.; Gerebtzoff, G.; Rodríguez-Pérez, R. Systematic evaluation of local and global machine learning models for the prediction of ADME properties. *Mol. Pharm.* **2023**, *20*, 1758–1767. [CrossRef] [PubMed]
61. Grebner, C.; Matter, H.; Kofink, D.; Wenzel, J.; Schmidt, F.; Hessler, G. Application of deep neural network models in drug discovery programs. *ChemMedChem* **2021**, *16*, 3772–3786. [CrossRef]
62. Narimatsu, S.; Arai, T.; Watanabe, T.; Masubuchi, Y.; Horie, T.; Suzuki, T.; Ishikawa, T.; Tsutsui, M.; Kumagai, Y.; Cho, A.K. Covalent binding of a reactive metabolite derived from propranolol and its active metabolite 4-hydroxypropranolol to hepatic microsomal proteins of the rat. *Chem. Res. Toxicol.* **1997**, *10*, 289–295. [CrossRef] [PubMed]
63. Oballa, R.M.; Truchon, J.F.; Bayly, C.I.; Chauret, N.; Day, S.; Crane, S.; Berthelette, C. A generally applicable method for assessing the electrophilicity and reactivity of diverse nitrile-containing compounds. *Bioorg. Med. Chem. Lett.* **2007**, *17*, 998–1002. [CrossRef]
64. Haduch, A.; Bromek, E.; Rysz, M.; Pukło, R.; Papp, M.; Gruca, P.; Łasoń, M.; Niemczyk, M.; Daniel, W.A. The effects of agomelatine and imipramine on liver cytochrome P450 during chronic mild stress (CMS) in the rat. *Pharmacol. Rep.* **2020**, *72*, 1271–1287. [CrossRef] [PubMed]
65. Masubuchi, Y.; Igarashi, S.; Suzuki, T.; Horie, T.; Narimatsu, S.J. Imipramine-induced inactivation of a cytochrome P450 2D enzyme in rat liver microsomes: In relation to covalent binding of its reactive intermediate. *J. Pharmacol. Exp. Ther.* **1996**, *279*, 724–731. [PubMed]
66. Walgren, J.L.; Mitchell, M.D.; Thompson, D.C. Role of metabolism in drug-induced idiosyncratic hepatotoxicity. *Crit. Rev. Toxicol.* **2005**, *35*, 325–361. [CrossRef] [PubMed]
67. Tettey, J.N.; Maggs, J.L.; Rapeport, W.G.; Pirmohamed, M.; Park, B.K. Enzyme-induction dependent bioactivation of troglitazone and troglitazone quinone in vivo. *Chem. Res. Toxicol.* **2001**, *14*, 965–974. [CrossRef]
68. Kassahun, K.; Pearson, P.G.; Tang, W.; McIntosh, I.; Leung, K.; Elmore, C.; Dean, D.; Wang, R.; Doss, G.; Baillie, T.A. Studies on the metabolism of troglitazone to reactive intermediates in vitro and in vivo. Evidence for novel biotransformation pathways involving quinone methide formation and thiazolidinedione ring scission. *Chem. Res. Toxicol.* **2001**, *14*, 62–70. [CrossRef]
69. Smith, M.T. Mechanisms of troglitazone hepatotoxicity. *Chem. Res. Toxicol.* **2003**, *16*, 679–687. [CrossRef] [PubMed]
70. Estupiñán, H.Y.; Wang, Q.; Berglöf, A.; Schaafsma, G.C.P.; Shi, Y.; Zhou, L.; Mohammad, D.K.; Yu, L.; Vihinen, M.; Zain, R.; et al. BTK gatekeeper residue variation combined with cysteine 481 substitution causes super-resistance to irreversible inhibitors acalabrutinib, ibrutinib and zanubrutinib. *Leukemia* **2021**, *35*, 1317–1329. [CrossRef] [PubMed]

biomolecules

MDPI

Article

Structural and Functional Characterization of Lipoxygenases from Diatoms by Bioinformatics and Modelling Studies

Deborah Giordano [1], Simone Bonora [1], Ilenia D'Orsi [1], Domenico D'Alelio [2,3] and Angelo Facchiano [1,*]

1 Institute of Food Sciences, National Research Council, via Roma 64, 83100 Avellino, Italy; deborah.giordano@isa.cnr.it (D.G.); sbonora@unisa.it (S.B.); ilenia.dorsi1990@gmail.com (I.D.)
2 Department of Integrative Marine Ecology, Stazione Zoologica Anton Dohrn, Villa Comunale, 80121 Naples, Italy; domenico.dalelio@szn.it
3 National Biodiversity Future Center (NBFC), Piazza Marina 61, 90133 Palermo, Italy
* Correspondence: angelo.facchiano@isa.cnr.it

Abstract: Lipoxygenases make several biological functions in cells, based on the products of the catalyzed reactions. In diatoms, microalgae ubiquitous in aquatic ecosystems, lipoxygenases have been noted for the oxygenation of fatty acids with the production of oxylipins, which are involved in many physiological and pathological processes in marine organisms. The interest in diatoms' lipoxygenases and oxylipins has increased due to their possible biotechnological applications, ranging from ecology to medicine. We investigated using bioinformatics and molecular docking tools the lipoxygenases of diatoms and the possible interaction with substrates. A large-scale analysis of sequence resources allowed us to retrieve 45 sequences of lipoxygenases from diatoms. We compared and analyzed the sequences by multiple alignments and phylogenetic trees, suggesting the possible clustering in phylogenetic groups. Then, we modelled the 3D structure of representative enzymes from the different groups and investigated in detail the structural and functional properties by docking simulations with possible substrates. The results allowed us to propose a classification of the lipoxygenases from diatoms based on their sequence features, which may be reflected in specific structural differences and possible substrate specificity.

Keywords: lipoxygenase; diatoms; protein modelling; sequence clustering; docking analysis

1. Introduction

Diatoms, microalgae ubiquitous in aquatic ecosystems, from freshwater to marine environments, are single-celled organisms with silica shells and multiple biotechnological purposes [1]. These microscopic algae are integral to aquatic food webs, supporting a myriad of organisms, generating oxygen, shaping food webs, and sequestering carbon [2]. The biological diversity and ecological adaptability of diatoms underscore their profound impact on aquatic ecosystems. The main driver of such an impact is the production of chemical compounds with extra-cellular functions, such as the oxylipins produced by lipoxygenase biochemistry.

Diatom-derived oxylipins are end products of well-characterized enzymatic pathways activated after cell wounding, starting from the lipolytic release of fatty acids (FAs) from complex lipids and proceeding through oxygenation of FAs by lipoxygenases (LOX) [3–8]. The ecological significance of diatom lipoxygenases and oxylipins extends beyond cellular metabolism, shaping community structure and functioning, highlighting their intricate role in maintaining the balance and resilience of aquatic environments. Oxylipins have multifaceted functions in aquatic ecosystems, mediating prey–prey and prey–predator signaling processes and communication within plankton food webs [9–11]. By producing oxylipins, diatoms influence the food searching behavior [12], reproductive fitness [13,14] and gene expression [15,16] of crustacean copepods, the dominant zooplankton assemblage that provides a strong trophic link between primary producers and secondary consumers (fish, jellyfish) [17]. In addition, there is interest in the possible biotechnological applications of diatom oxylipins for exploiting, as some studies suggested, their anti-cancer, anti-bacterial, anti-fungal and anti-parasitic activities (for a review see [18].

Besides their role in aquatic environments as info-chemicals, allelochemicals, deterrents against grazers and mediators influencing carbon recycling, recent studies additionally demonstrated that oxylipins exert anti-inflammatory activity, can promote tissue regeneration, alleviate pain [19], and regulate angiogenesis, blood pressure, blood coagulation, and blood vessel permeability [20]. Moreover, due to their capability to promote the activation of apoptotic pathways in cancer cells, to regulate adhesion, migration and proliferation, oxylipins can also exert antiproliferative activity on human cancer cells [21–23].

The LOX enzymes are responsible for the oxidation of polinsatured fatty acids to oxylipins. This reaction takes place in the LOX catalytic U-shaped pocket in the presence of an iron ion as a cofactor and a water molecule. The cofactor is located in a central position, at the base of the "U pocket", and is usually coordinated by five residues: three His, one Asn and one Ile. The iron cofactor extracts a hydrogen atom to the nearest carbon of the substrate, at two positions upstream or downstream of the oxidation of the double bond, according to the substrate orientation in the pocket [24]. Generally, if the substrate polar head is located externally to the active site, the oxidation takes place two positions after the carbon atom nearest to the iron; otherwise, if the tail is oriented to the external side, then it takes place two positions before, based on the substrate double bond placed at a minor distance from the oxygen channel present in the protein. According to the position of the double bond oxidized, it is possible to have different LOX types also for the same substrate [25]. The LOX 3D structure is generally well preserved among bacteria, plants and mammals [26], however, the amino acid composition of the U-shaped cavity jointly with the conformational change that could be induced by substrates or inhibitors, plays a crucial role in LOX specificity [27]. A detailed explanation about the enzyme substrate recognition mechanism and about the enzyme activity can be found in ref. [24] and [27] and in their figures. Moreover, a visual representation of the interaction occurring among substrate cofactor and coordination residues is visible in Supplementary Figure S1.

In this study, we analyzed a wide range of biomolecular data relating to diatom LOXs available from public repositories, from which we modelled the fine-scale 3D structure of different categories of these enzymes present in the biochemical machinery of diatoms belonging to different phylogenetic groups. The detailed structural and functional analysis suggested a classification and a characterization of the putative LOX from diatoms, giving a hypothesis on their substrate specificity.

2. Materials and Methods

2.1. LOX Sequences: Detection, Analysis and Clustering Procedure

Starting from the sequences annotated as LOX in Uniprot and NCBI, it was possible to select two reference sequences: the LOX from *Fragilariopsis cylindrus* CCMP1102 (accession number NCBI: OEU18547.1); and the LOX from *Pseudo-nitzschia arenysensis* (accession number NCBI: QWC64745.1). These sequences have been used as NCBI BLAST [28,29] queries to search for other putative LOX protein sequences in diatoms by exploiting different databases of both protein sequences and transcripts. In the details, blastp was used to search the two queries versus the Non-redundant protein sequence (nr) database and the Transcriptome Shotgun Assembly proteins (tsa nr) database in the taxon of diatoms (taxid: 2836); while tblasn was exploited for searching the queries versus the Nucleotide collection (nr/nt) database, the Whole-genome shotgun contigs (wgs) database, and the Transcriptome Shotgun Assembly (TSA), again, narrowing the search to the taxid: 2836. The hypothetical LOX nucleotide sequences obtained have been translated by the Expasy Translate Tool [30].

Similarly to the approach used in previous studies of our lab [31], sequences collected have been standardized by performing multiple sequence alignments using Clustal Omega [32] by a visual inspection, focusing on their length and the presence along the sequence of the LOX domain, and by manually cutting and removing each different domain detected. Domain detection has been performed by exploiting the ScanProsite tool [33] and Search InterProScan [34]. The standardized sequences obtained have been clustered by the construction of several phylogenetic trees performed by MEGA tool 6.06 [35], using Maximum Likelihood algorithm, a bootstrap of 100, the Jones–Taylor–Thornton (JTT) substitution model, all sites for gaps data treatment, and the Subtree-Pruning-Regrafting-Fast (SPR level 3) for the ML heuristic method. Group and subgroup division were performed on the base of the obtained branches, grouped by a bootstrap value major of 95 and according to the mutation occurring at the fourth cofactor coordination residue.

To search conserved regions, motifs or patterns along the sequences of each group detected after the clustering procedure, MEME suite [36] was used.

2.2. 3D Models Construction and Docking Analysis

For each group detected, a representative protein sequence was chosen for further structural and docking analyses. Their simple modeling by AlphaFold2 was not satisfactory because the procedure did not generate the active conformation of the enzyme, with iron ion and related coordinating side chains correctly folded, functional water molecule, and appropriate access of the substrate to the active site. Therefore, we adopted a strategy to model LOXs of *Fragilariopsis cylindrus* and *Pseudo-nitzschia arenysensis* with the correct active conformation features, and then used these models as a template for modelling the other LOX sequences. Similar procedures are routinely applied in our laboratory for several protein models' construction and investigation (for a recent example, see [37]). In detail, the crystallographic structure of the allene oxide synthase-lipoxygenase protein from *Plexaura homomalla* (PDB code: 3FG1) and of the lipoxygenase 3 from *Glycine max* (PDB code: 1HU9) have been used as templates for modelling the LOX sequences isolated from *Fragilariopsis cylindrus* and *Pseudo-nitzschia arenysensis*, respectively. A first multiple modelling step was performed using SWISS-MODEL [38]; subsequently, the best model obtained for each sequence was used, jointly with the PDB structures mentioned before, as template for the building of the final models employing Modeller10.2 [39]. Each model obtained is subjected to a validation process that considers the stereochemical model quality, its total energy, its geometrical score and the most suitable conformation (correct position of the cofactor coordination atoms, open conformation of the active site, etc.). Conformational energy, backbone and side-chain stereo-chemical characteristics, and suitability with known properties were evaluated using the web servers ProSA-web [40], QMEAN [41] and PROCHECK [42].

The best models obtained have been exploited for the model refinement of the Alpha Fold models obtained for the other sequences selected. Therefore, the two models of LOX from *Fragilariopsis cylindrus* and *Pseudo-nitzschia arenysensis* were in turn used as templates by Modeller10.2 to refine the models of the other representative putative LOX sequences built by the ColabFold tool of ChimeraX v1.5 [43]. Details about the modelled sequences and their templates are reported in Table 1.

This refinement procedure allowed us to construct all models in active conformation, in presence of an iron ion as a cofactor, with a correct orientation of its coordination residues, and of a functional water molecule; the latter added exploiting the 3D structure of a lipoxygenase in a complex with arachidonate from *Plexaura homomalla* (PDB code: 4QWT).

Once created the 3D structures in the active conformation of all the sequences selected, focused docking analyses with flexible and rigid protocols were performed using AutoDock and AutoDockTools 4.2.6. [44], testing as substrates: eicosapentaenoic acid (EPA) (PubChem CID: 446284), hexadeca-6Z,9Z,12Z-trienoic acid (HTrA) (PubChem CID: 52921838), hexadeca-6Z,9Z,12Z,15-tetraenoic acid (HTtA) (PubChem CID: 11957734), docosa-4Z,7Z,10Z,13Z,16Z,19Z-esenoic acid (DHA) (PubChem CID: 445580), 9,12-hexadecadienoic acid (PubChem CID: 5282787), and 4,7,10,13,16,19,22,25-octacosaoctaenoic acid (28: 8n–3) (PubChem CID: 54316623). Ligands were made flexible in all simulations.

To allow the accommodation of the substrates, flexibility in some amino acid side chains was adopted when some pockets, due to the particular hindrance of some amino acids that composed them, did not allow substrate binding by a simple rigid methodology of docking. Details about the docking parameters adopted are reported in Supplementary Table S1. As positive control of the procedure, a re-docking was simulated exploiting the crystallographic structure of the complex made up of the LOX from *P. homomalla* and the arachidonic acid (AA) (PDB code: 4QWT). The binding energy values we report refer to the free energy of binding of ligands to receptors estimated by the AutoDock 4.2 force field.

Table 1. Reference sequences and their templates. The table reports information related to the NCBI accession number of the putative LOX sequences, their annotation, the organism of origin, and the two templates used for modelling performed by Modeller10.2.

NCBI Accession Number	Annotation	Organism	Template 1	Template 2	Sequence Identity
OEU18547.1	lipoxygenase	*Fragilariopsis cylindrus*	SWISS-MODEL model	3FG1	20.88%
QWC64745.1	lipoxygenase	*Pseudo-nitzschia arenysensis*	SWISS-MODEL model	1HU9	26.64%
CAD9618313.1	unnamed protein product	*Skeletonema marinoi*	AlphaFold model	LOX from *P. arenysensis* model	45.32%
XP_002180840.1	predicted protein	*Phaeodactylum tricornutum*	AlphaFold model	LOX from *P. arenysensis* model	32.10%
CAD8595700.1	unnamed protein product	*Asterionellopsis glacialis*	AlphaFold model	LOX from *F. cylindrus* model	26.22%
GJWO01009215.1	transcribed RNA sequence	*Bellerochea* sp.	AlphaFold model	LOX from *F. cylindrus* model	22.62%
CAE0460244.1	unnamed protein product	*Chaetoceros debilis*	AlphaFold model	LOX from *F. cylindrus* model	23.61%
CAD8307630.1	unnamed protein product	*Pseudictyota dubia*	AlphaFold model	LOX from *P. arenysensis* model	22.30%

3. Results

3.1. Putative LOX Sequences Detection, Analysis and Classification

From sequence database searching, we found that the only protein sequences annotated as lipoxygenase in diatoms are the ones from *Pseudo-nitzschia arenysensis*, *Fragilariopsis Cylindrus*, and *Nitzschia inconspicua*. Although the first two have the same annotation and a similar length, 704 residues for LOX from *P. arenysensis* and 742 for LOX from *F. cylindrus*, from their alignment it is possible to notice that they are dissimilar; it is possible to detect a similarity of 21% and in only 20% of sequence coverage. Concerning the LOX of *N. incospicua*, this sequence seems to be wrongly annotated due to the absence of a domain attributable to LOX, and from its modelling the structure obtained is not comparable to a typical LOX structure. Thus, starting from these two sequences, different transcriptome and protein sequence databases have been searched (see methods for details), leading to the detection of about a hundred sequences, most of them annotated as hypothetical proteins. Sequence fragments, or sequences not preserving at least three of the five cofactor coordination residues, and sequences without a putative LOX domain have been removed from the list.

By this procedure, it was possible to collect 45 possible LOXs from diatoms, found to be similar to the LOX from *P. arenysensis* and *F. Cylindrus* (Supplementary Table S2). Thus, starting from an optimized multiple alignment of all the sequences selected, a clustering analysis was performed by the construction of a phylogenetic tree. As expected, the results obtained highlight that the sequences found can be divided into two major groups: one collecting sequence, similar to LOX from *P. arenysensis*, and the other one from *F. Cylindrus*. However, the tree analysis suggests that the 45 sequences can be further divided into at least six groups (bootstrap value on branch grouping them > 95) (Figure 1A), among which, groups four and five can be in turn divided into two subgroups according to the mutation of the fourth cofactor coordination residue and the presence of two different branches grouping sequences with a bootstrap value equal to 100. Groups 1, 2, and 3 include most of the sequences similar to LOX from *P. arenysensis,* although a few other sequences are on the opposite side of the tree (group 6).

A

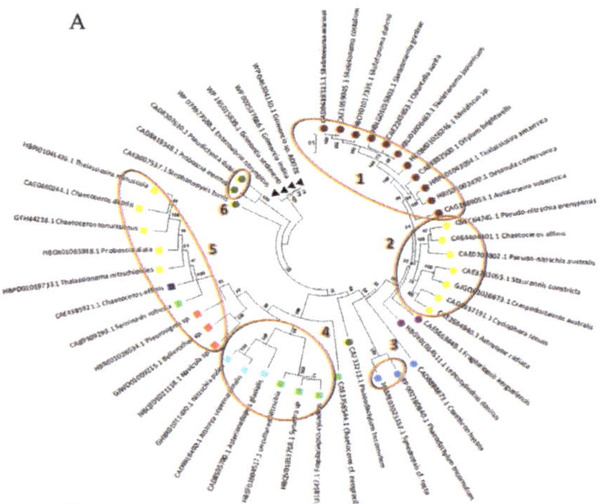

B

Group	Symbol	Representative organism	Iron coordination site
1	●	*Skeletonema marinoi*	HLG[KNEDR]TH and HAAVN
2	●	*Pseudo-nitzschia arenysensis*	HL[GA][MKNR]TH and HAAVN
3	●	*Phaeodactylum tricornutum*	HSWVH and HSTDH
4a	■	*Fragilariopsis cylindrus*	HLLE[IV]H and HN[LV]VAN
4b	■	*Asterionellopsis glacialis*	HL[GV][HN][VC]H and HE[LH]VGQ
5a	■	*Bellerochea sp.*	HL[FLY][NW][ACL]H and H[ED][HLQ]VGS
5b	■	*Chaetoceros debilis*	HL[DACI][WQ][TS]H and HE[IL][MIV]GH
6	●	*Pseudictyota dubia*	HF[IV][EY][TS]H and HH[VS][MW]N

Figure 1. Phylogenetic analysis. (**A**) Phylogenetic tree representing the clustering of the LOX sequences analyzed. Highlighted by number and circle are the six different major groups detected. Marked by a circle are the sequences similar to LOX from *P. arenysensis*, and by a square the ones more similar to *F. cylindrus*. In group 4, under the same branch grouping them with a bootstrap value of 100, subgroups can be distinguished as 4a (marked by green squares) and 4b (by cyan squares). Similarly, in group 5, subgroup 5a (red square under the same branch) is distinguishable from the rest (group 5b under the branch with a bootstrap value of 100) due to the same bootstrap split; LOX sequence from *Pleurosigma* sp. seems to be an intermediate form between the two subgroups. Black triangles indicate sequences exploited for the root of the tree. The colors of the labels refer to specific iron coordination residue compositions visible in the symbol section reported in panel B. (**B**) Reference sequence representative of each group detected by the phylogenetic three and their relative cofactor coordination pattern. Each group has as a symbol: a circle or a square, according to the major similarity with LOX from *P. arenysensis* or *F. cylindrus*, respectively. Different colors refer to specific labelled sequences grouped by a circle in the tree. Residues directly involved in the cofactor coordination are highlighted in bold.

All groups, except for the first and the second, differ either in the distance along the sequence of the second couple of cofactor coordination residues, or in the fourth coordination residue that could be histidine, glutamine, or serine, instead of asparagine.

In detail, group 3 presents a deletion between the first two histidine residues of coordination, and the substitution of the asparagine of coordination with another histidine. Group 4 presents, in both its subgroups, an insertion between the third and the fourth coordination residue; however, in subgroup 4b this residue is substituted by glutamine. Group 5 presents an insertion between the third and the fourth coordination residue too, but its asparagine of iron coordination is substituted by serine and histidine in 70% of the sequences composing subgroups 5a (yellow squares in the tree) and all the sequences composing subgroup 5b plus LOX from *Pleurosigma* sp. (red squares in the tree), respectively. Group 6 seems to preserve, similar to groups 1 and 2, the same cofactor coordination residues; however, the position of the third residue represents a tricky point due to the presence of two histidine residues one after the other that do not allow to detect which one is involved in the coordination, on the basis of sequence alignment. A detailed description of the coordination pattern is displayed in Figure 1B (last column).

From the analysis of motifs/patterns present along the sequences, it was possible to detect the reason why groups 1 and 2 emerge as separated. Even if both groups preserve the same coordination residues, they differ in the presence of a specific pattern in group 2 that is absent in group 1 and vice versa (Supplementary Figure S2).

3.2. 3D Models and Docking Analysis

For each group, we chose a representative protein sequence that we modelled in an active conformation, in the presence of both water and a cofactor. For the first group, we selected LOX from *S. marinoi*; for the second, LOX from *P. arenysensis*; for the third, LOX from *P. tricornutu*, LOX from *F. cylindrus*, and *A. glacialis* for groups 4a and 4b, respectively; LOX from *Bellerochea* sp. for group 5a and LOX from *C. debilis* for 5b; and LOX from *P. dubia* for group 6 (Figure 1B).

From the models obtained it was possible to notice that the deletion that occurred in group 3 does not seem to affect the correct orientation of the coordination residues (Supplementary Figure S3C). In a similar way, in groups 4 and 5, despite the insertion, the asparagine or the residue which replaces it, i.e., glutamine, histidine or serine, is able to reach the correct position to coordinate the cofactor (Supplementary Figure S3D–G). Moreover, the model obtained for the sequence representative of group 6 highlighted that between the two consecutive residues of histidine found in the correct position for the cofactor coordination, the functional role seems to be carried out by the first (Supplementary Figure S3H). After checking the quality of the models obtained and the compatibility of their structures with an active conformation, docking analyses were performed with EPA, the main fatty acid present in diatoms, or other suitable substrates selected from the literature, to analyze the LOX type to which each putative LOX selected belongs to.

In the quality of positive control of the docking approach applied, the first analysis performed regarded the re-docking of the LOX from *P. homomalla* and the arachidonic acid; complex is already available in the PDB. Our results completely reproduced the crystallographic data and as reported in the literature [27], indicating LOX8-type activity. As in the crystallographic structure, the AA placed itself with its polar head outside the catalytic pocket, and the cofactor interacts with the C10 of the substrate. Due to the presence of a Gly in position 427 that defines the top of the putative O2 channel, the oxygen necessary for the reaction is free to flow in this upper side of the channel [45,46] and to attack the C8 of the AA, leading to the generation of 8R products.

Similarly, we have analyzed docking results obtained for the other representative putative LOXs with the suitable substrates selected. Results have shown that the pocket composition determines different interactions for the same substrate and influences the substrate selectivity of the enzyme (Table 2).

Table 2. Most relevant docking results. The binding energy column indicates the medium binding energy calculated from the mean of all the binding energies detected for each binding pose belonging to the cluster in exam. The acronym l.b.e reported in round brackets stands for lowest binding energy and is related to the pose in the cluster that assumes the lowest energy of interaction.

Group	Organism of Origin	Substrate	Binding Energy (Kcal/mol)	N° of Poses in Cluster	Ki (μM)	Residues of Interaction (H-bonds are underlined)	LOX Type
control	*Plexaura homomalla*	AA	−6.95 (l.b.e.= −7.40)	15	3.74	Tyr178-Arg182-Gln380-His384-Leu385-His389-Gly427-Leu431-Ser441-Leu444-Thr623-Val624-Leu627-Fe^{2+}-H_2O	LOX8
1	*Skeletonema marinoi*	EPA	−5.46 (l.b.e.= −7.04)	11	6.93	Gln266-Trp267-His270-Leu271-His275-<u>Ser317</u>-Leu318-Phe326-Gln330-Met529-Leu533-Ile599	LOX9
		HTrA	−5.52 (l.b.e.= −5.79)	11	56.56	Ala213-<u>Gln266</u>-Leu271-His275-Ala314-Leu318-Phe326-Gln330-<u>Ser526</u>-Met529-Val530-Leu533	LOX9
		HTrA	−4.89 (l.b.e.= −5.46)	4	334.45	Gln266-His270-Leu271-His275-Ala313-Ala314-Ser317-Leu318-Phe326-Gln330-<u>Met529</u>-Leu533	LOX9
2	*Pseudo-nitzschia arenysensis*	EPA	−7.59 (l.b.e.= −7.59)	1	2.75	Glu388-Trp389-Leu393-Leu432-Thr439-Leu440-Phe448-Gln583-Ile634-Ile637-Leu638-Ile704-Fe^{2+}-H_2O	LOX12
3	*Phaeodactylum tricornutum*	9,12-hexa-decadienoic acid	−5.62	1	75.75	Ala97-His101-His105-Ser102-Phe352-Ala353-Phe356-Val357-Ala147-Val150-Gln151-Phe198-Trp418-Fe^{2+}	LOX12
4a	*Fragilariopsis cylindrus*	EPA	−5.63 (l.b.e.= −5.81)	2	54.97	Phe160-<u>His164</u>-Leu165-Ile203-Ala207-Leu211-Ala355-Asn356-Val397-Leu400-Thr401-Fe^{2+}-H_2O	LOX8
			−5.51	1	91.52	Phe160-His164-Leu165-His169-Ile203-Ala207-Leu211-Ile218-Thr221-Val397-Thr401-Leu478-Fe^{2+}	LOX15
4b	*Asterionellopsis glacialis*	EPA	−5.25	1	141.17	Thr459-His463-Leu464-Val503-Ser506-Ala507-Ile510-Leu511-Phe517-Ile518-Trp522-Leu663-Leu690-Thr697-Fe^{2+}	LOX9
		DHA	−4.77 (l.b.e.= −5.60)	4	78.95	His463-Leu464-Leu511-Phe517-Ile518-Trp522-Leu690-Leu693-Val694-Ser696-Thr697-Leu77-Fe^{2+}	LOX14
		Octacosa-octaenoic acid	−6.50 (l.b.e.= −6.93)	3	17.41	Phe-455-Thr459-His463-Leu464-Val503-Ser506-Ala507-Leu511-Phe517-Ile518-Val521-Trp522-Leu690-Leu693-Val694-Thr697-Ser766-Leu767	LOX20
5a	*Bellerochea* sp.	DHA	−5.20 (l.b.e.= −6.29)	6	25.02	His279-Thr275-Leu280-Thr322-Ala323-Ala326-Leu327-Leu333-Phe337-Glu461-Thr471-Ser488-Leu491-Met492-Leu494-Val558-H_2O	LOX11
		Octacosa-octaenoic acid	−5.80 (l.b.e.= −7.51)	6	3.14	Val271-Thr275-Ile319-Thr322-Ala323-Leu327-Leu333-Glu461-Thr464-Ser465-Ala469-Gly470-Trp485-Ser488-Leu491-Met492-Leu494-Val558-Fe^{2+}-H_2O	LOX11
5b	*Chaetoceros debilis*	EPA	−5.78 (l.b.e.= −5.87)	6	49.54	Thr337-His341-Leu342-His346-Ser388-Leu389-Thr395-Ile396-Ala399-Phe538-Ile559-Thr562-Thr563-Fe^{2+}-H_2O	LOX9

Table 2. *Cont.*

Group	Organism of Origin	Substrate	Binding Energy (Kcal/mol)	N° of Poses in Cluster	Ki (μM)	Residues of Interaction (H-bonds are underlined)	LOX Type
6	*Pseudictyota dubia*	EPA	−5.64 (l.b.e.= −5.89)	9	47.91	Gln256-Met257-Phe260-His264-Asn298-Gly302-Leu306-Tyr313-Phe482-Leu483-Phe486-Phe487-H_2O	LOX8
		DHA	−6.55 (l.b.e.= −7.35)	6	4.08	Phe82-Gln256-Phe260-His264-Asn298-Ala301-Gly302-Leu306-Tyr313-Trp317-Gly439-Ser443-Ile444-Leu483-Thr484-Phe487	LOX4/7
		Octacosa-octaenoic acid	−6.53 (l.b.e.= −7.97)	19	1.43	Phe82-Gln256-Met257-Phe260-His264-His259-Asn298-Ala301-Gly305-Leu306-Thr311-Thr312-Tyr313-Trp317-Leu483-Phe487	LOX16

3.3. Docking Analyses for Sequences Similar to LOX from P. arenysensis

For LOX from *P. arenysensis*, docking with EPA has shown an interaction similar to that of the positive control. For LOX from *P. homomalla*, six conserved amino acids, i.e., the first coordination couple (His384/389), three Leu (385/431/627) and one Ile (626), are strategically located to allow the attack of the iron ion on the C10 of the substrate [27] (Figure 2A); all these residues are conserved also in the LOX from *P. arenysensis*.

According to our results, the EPA carbon that interacts with the iron ion is carbon 10. However, in this case Ala (in position 436, in place of the corresponding Gly427 in LOX from *P. homomalla*), determines the occlusion of the upper side of the oxygen channel, driving the O2 deeper into the cavity; this causes an O2 reaction with the carbon 12 of the substrates; therefore, this enzyme shows a LOX12 behavior. Moreover, the binding energy and constant of inhibition are comparable between the LOX of *P. arenysensis* and the control (−7.40 Kcal/mol with a Ki of 3.74 μM for LOX from *P. homomalla* and −7.59 Kcal/mol with Ki equal to 2.75 μM for LOX from *P. arenysensis*). Interestingly, the LOX from the *P. homomalla* active site is characterized by a deep cavity due to the presence at its terminal part of two Alanine residues in position 589 and 620. The LOX from *P. arenysensis* does not preserve these residues, showing in their place Leu595 and Val631; however, this cavity reduction is not enough to impede EPA binding (Figure 2B).

In the case of LOX from *S. marinoi*, the active site preserves the six conserved amino acids involved in the correct location of the substrate, even if Ile626 is replaced by a leucine, but changes the composition of the terminal part of the pocket with Ala589/620 replaced by Ser526 and Trp474. The presence of Trp in place of Ala determines a shallower cavity, while the Ser is suitable for better interaction with the polar head of the substrates. From our simulations is possible to detect that EPA is not able to interact using its C10 with the iron ion, whose attack is limited to C7 (Figure 2C). This conformation results in a higher binding energy (mean binding energy equal to −5.46 Kcal/mol, lowest binding energy equal to −7.04 Kcal/mol) than the ones detected for the positive control (mean binding energy equal to −6.95 Kcal/mol, lowest binding energy equal to −7.40 Kcal/mol). The reduction of the pocket volume, due to the Trip presence, probably limits the deepening of the hydrocarbon tail. For what concerns the smaller substrates tested, HtrA and HTtA, both bind the active site with their carboxyl head located inside the pocket and interact with the iron ion by their C11. Also in this case, the upper side of the oxygen channel is locked by the presence of the Ala314, so the enzyme behavior can be associated with a LOX9 for all three substrates tested. Generally, looking at the binding energies, we can see that they are comparable for EPA and HTrA and higher for HTtA, revealing a preference of this LOX for EPA and HTrA.

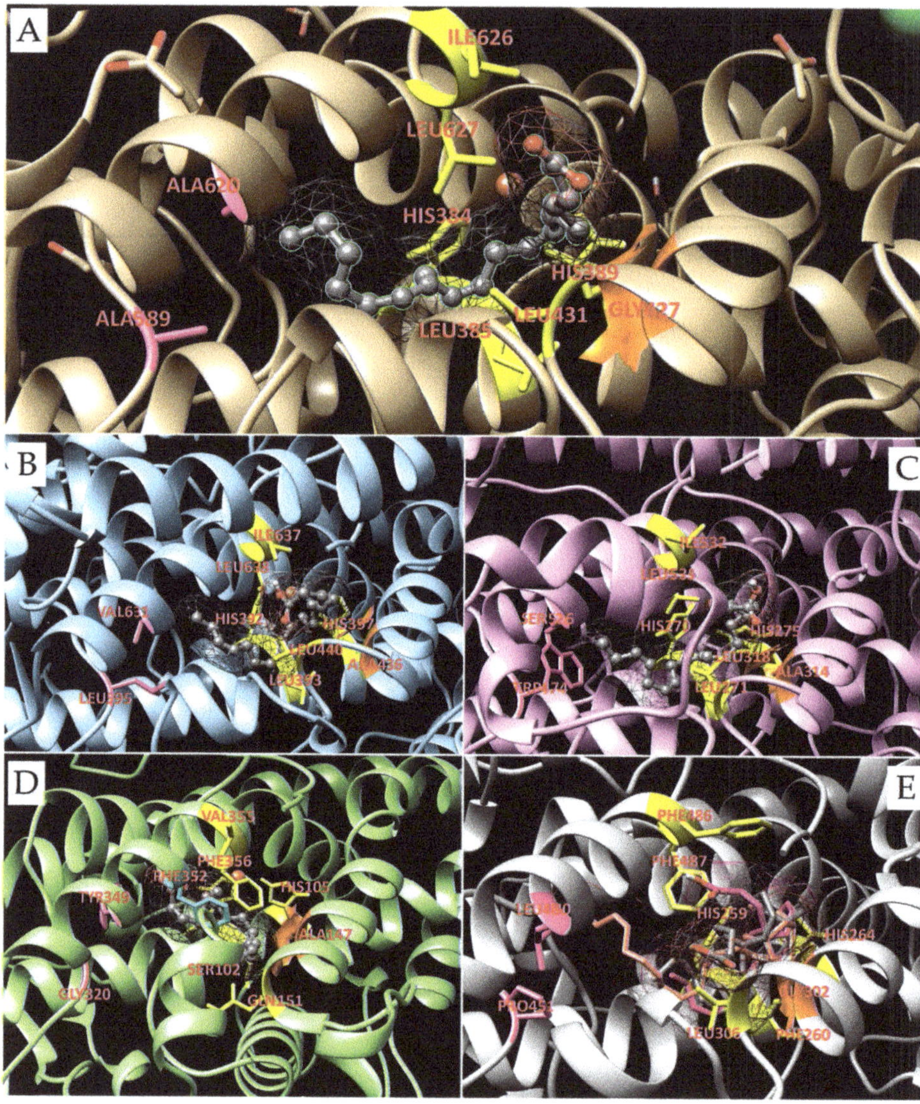

Figure 2. Active site pocket composition in LOX from *P. homomalla* (**A**), *P. arenysensis* (**B**), *S. marinoi* (**C**), *P. tricornutum* (**D**) and *P. dubia* (**E**). The ligands are represented with a grey ball and stick; the residues responsible for their orientation in the pocket are highlighted in yellow sticks; while the ones composing the deepest portion of the catalytic pocket are highlighted in pink; and glycine and alanine, that open or close, respectively, the upper side of the oxygen channel are represented in orange surfaces. In (**D**), represented in cyan sticks, Phe352 probably partially closes the deepest region of the active site; while in (**E**), beyond the octacosaoctaenoic acid, represented in grey sticks, DHA and EPA are shown in pink and magenta, respectively.

Hypothetical LOX from *P. tricornutum* presents many differences from the ones described until now. Of the six residues implicated in the orientation of the substrate, it preserves only the two histidines, while Leu385/431/627 is replaced by Ser102, Gln151 and Phe356, and Ile626 is replaced by Val355. Moreover, Ala589/620 is replaced by Gly320 and Tyr349, and Gly427 by Ala147. All these modifications influence the substrate binding, which seems to be able to move deeper inside the external part of the pocket, and Phe356 seems to drive the substrate fold inside it. Moreover, the substitution of Thr623 with Phe352 partially closes the deepest region of the active site, while the deletion between the two His of coordination and the replacement of Leu385 with Ser102 create more space in this area. Docking simulations performed with EPA for this hypothetical LOX did not highlight any noteworthy result; simulations performed with 9,12-hexa-decadienoic acid show a binding energy comparable to the one obtained for *S. marinoi* LOX and suggest a possible interaction of the iron ion with the C8, resulting in a possible LOX12 because of the closure of the upper side of the oxygen channel. O2 is probably able to reach only the last part of the ligand thanks to its particular orientation. The substrate seems capable of binding the pocket in an unconventional manner, with the polar head and the tail oriented toward Val355 and Gln151, respectively (Figure 2D). Among the representative sequences selected, the last one, more similar to the one from *P. arenysensis*, is the hypothetical LOX from *P. dubia*. This sequence preserves the coordination residues and the canonical number of residues among them; however, it presents a higher variability in the composition of this region. Many substitutions occur; in particular, Pro451 and Leu480 completely close the end of the pocket, replacing Ala589/620; among the six residues necessary for locking the substrate conformation suitable for the iron attack, only His384/389 and Leu431 (at positions 259/264 and 306, respectively, in this sequence) are preserved, the others are replaced by three phenylalanine residues. The Gly427 (Gly302 in LOX from *P. dubia*) opening the higher portion of the oxygen channel is preserved. Docking simulations have been performed with different substrates: EPA, DHA and octacosaoctaenoic acid. All the ligands are positioned with their hydrocarbon tail inside the deepest portion of the pocket and their carboxyl head outside (Figure 2E). However, each substrate binds the iron ion at different positions of their hydrocarbon chain: EPA with its C10, DHA with C6 or C9 that are equally distant from the iron, even if looking at the substrate conformation the interaction with the C9 seems the most probable, and octacosaoctaenoic acid with C18, resulting in a LOX8, LOX4/7 or LOX16, respectively. Looking at the binding energy, it is possible to detect a preference for DHA and octacosaoctaenoic acid, which show the lowest binding energies and Ki values comparable to the ones detected for AA and EPA interaction with the LOX from *P. homomalla* and *P. arenysensis*, respectively.

3.4. Docking Analyses for Sequences Similar to LOX from *F. cylindrus*

Docking results related to LOX from *Fragilariopsis cylindrus* are a clear example, as the different orientations of the substrates determine different LOX types. The structure preserves four of the six amino acids necessary for the correct substrate orientation in the pocket, with Ile626 and Leu627 replaced by Leu664 and Thr665, respectively. Ala471 locks the upper side of the O2 channel. Gln658 and Gly633 replace the Ala589/620 forming the back of the pocket. Moreover, in this region, Gln380 and Val438 present in LOX from *P. homomalla* are replaced by Phe424 and Ile482, which shrink the area. In particular, Phe424 opposite to Thr485, jointly with Gln658, further reduces the access to the back of the pocket. The second coordination couple, i.e., His615 and Asn620, preserves its correct conformation for the interaction with the iron ion. However, the insertion between the couple of one more residue, in addition to the canonical three, results in the break of the alpha helix hosting them and determines a protrusion in the consequently formed coil inside the active site pocket. Moreover, the presence of Thr623, which replaces the Gln577 of LOX from *P. homomalla*, residue preserved also in LOX from *P. arenysensis* and *S. marinoi*, together with the already mentioned Thr665 in place of Leu627, and Phe424 in place of Gln380, creates a larger area before the access to the deeper portion of the cavity. This region is comfortable for the substrate docking, which could make novel interactions with the portion embedded between the coordination residue couple His/Asn, including also the backbone of the novel inserted Ala619. Thus, this insertion, jointly with the substitution that occurred in the deeper portion of the active site, results in a slight modification of the active site pocket shape that becomes larger but shorter (Figure 3A,B) and is suitable to host the substrate in different conformations.

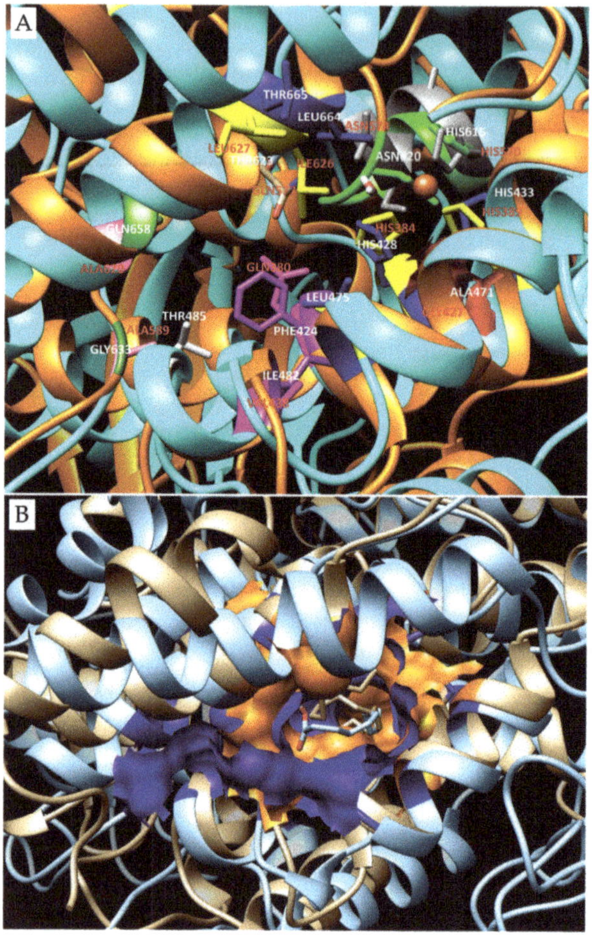

Figure 3. Active site pocket comparison between LOX from *F. cylindrus* and LOX from *P. homomalla* or/LOX from *P. arenysensis*. (**A**) Comparison of the aminoacidic composition of the active site pockets present in LOX from *P. homomalla* and *F. cylindrus*. The superimposition procedure was adopted, but the two sequences are completely different, and this has an impact on the two structures whose overlap generates a high RMSD value, i.e., 4.615 Å. In orange ribbons are the LOX from *F. cylindrus* with relevant residues composing the active site pocket highlighted in blue, violet, green, white, dark green and cyan sticks, all indicated by white labels; while the Ala471 locking the O2 channel is represented with a red surface. The LOX from *P. homomalla* is represented in orange cartoons with its relevant residues highlighted in yellow, pink, magenta, orange, grey and beige, all indicated by red labels. (**B**) Shape comparison of the active site pocket of LOX from *P. arenysensis*, with a blue surface in a structure represented in cyan ribbons; and LOX from *F. cylindrus*, with an orange surface in a structure represented in beige ribbons. In this case, too, the superimposition of the structures has a high RMSD value, i.e., 8.866 Å, due to very strong difference between the protein sequences. The different active site compositions determine differences in terms of pocket shape and substrate binding.

Our docking results show that EPA can bind the active site in a double conformation, for each of which its carboxyl head is located either inside the larger area at the entrance of the deeper portion of the pocket or in the first part of the latter. The EPA tail, instead, is oriented either in the external part of the pocket or folded, exploiting the external part of the pocket at the larger area in the center. In particular, when the polar head of EPA is oriented in the center of the pocket, the iron interacts with the carbon 10 of the substrate, and vice versa; when it is oriented in the deeper portion of the pocket, the carbon nearest to iron, is C13. This means that in the first case, the enzyme behavior is typical of a LOX8 in the second of a LOX15 (Figure 4). No preference seems to be present for one or the other conformation because both the binding energies are quite comparable, as are the inhibition constants.

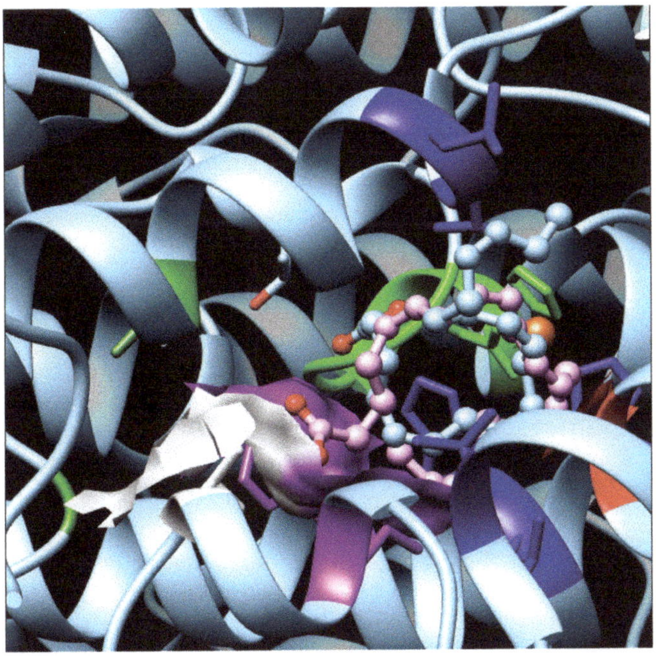

Figure 4. Different orientations of the EPA substrate in the LOX from the *F. cylindrus* active site. The enzyme is represented by a cyan carton; Phe424 and Ile482, which narrowed the entrance in the deepest portion of the cavity, are highlighted in a violet surface together with Thr485 in white surface; in the blue sticks are the residues responsible for the correct orientation of the substrate in the pocket; in green is the region between the second couple of iron coordination residues: His615 and Asn620 are highlighted with a stick. In cyan is Thr623, which creates further space in the cavity. EPA is represented with pink/cyan balls and sticks.

A clustering analysis collected the sequences of the group represented by *F. cylindrus* LOX sequence near another one, for which LOX from *Asterionellopsis glacialis* has been chosen, but under the same clade. A hypothetical LOX from *A. glacialis* preserves four residues of the six necessary for the ligand orientation. In particular, Ser696 and Thr697 substitute Ile626 and Leu627, respectively; the deepest portion of the pocket is composed of Thr673 and Leu690 in place of Ala589/620, and Gln380 and Val438 (detectable in LOX from *P. homomalla*) are replaced by Thr459 and Phe517, respectively; while the upper side of the O2 channel is closed by Ala507 instead of Gly427. The presence of Gln660 in the replacement of the Asn of coordination seems to not affect the iron binding; however, in this case also, the insertion among the last couple of coordination residues breaks the alpha helix. Having no information about the enzyme products in this organism, for this LOX, three different substrates with different lengths have been chosen to detect a possible preference. All the substrates bind the enzyme with their polar head outside and their tail inside the pocket; however, the EPA bound the iron with its C6. Therefore, the oxidation occurs in position 9, DHA with its C12; thus, the enzyme behavior is typical of a LOX14, instead of the octacosaoctaenoic acid with its C18, so the oxidation happens in position 20.

In group 5, two main clades can be distinguished, so we decided to analyze them both, focusing our attention primarily on the hypothetical LOX sequence from *Bellerochea* sp., which seems drastically split from the other sequences. Among these, we selected hypothetical LOX from *Chaetoceros debilis* that appears to be the most recent.

Hypothetical LOX from *Bellerochea* sp. seems to be characterized by the deepest pocket. Moreover, Ala620 is preserved while Ala589 is substituted by Thr471, but the residues preceding it, Ala-Pro in *P. homomalla* LOX, are replaced by Ala and Gly in *Bellerochea* sp. Moreover, residues from 579 to 583 (Asp-His-Tyr-Gly-Phe), in alpha helix conformation in *P. homomalla*, are replaced by a coil (Leu-Thr-Ser-Pro), which creates more space in the last portion of the pocket and reduces the central part of it. That is because the iron coordination residues His453 and Ser458 are found just before these residues, and the insertion among this couple generates, as in the previously mentioned LOX, a break in the helix and a kinked conformation of the coil that invades the active site and warps it. Among the residues of coordination of the substrate, His279/284 and Leu280/327 (corresponding to His384/389 and Leu385/431 in LOX from *P. homomalla*, respectively) are preserved; Leu494 and Thr495 replace Ile626 and Leu627. Also in this case, the oxygen channel is closed by the replacement of Gly427 by Ala323, so the oxidation occurs at the deepest region of the ligand, for which the double bond is located closer to the iron. Docking performed with HTrA and EPA have not detected good results, while DHA and octacosaoctaenoic acid have shown a good capability to bind the active site, suggesting a preference for a bigger ligand. In both cases, the tails reach the deepest portion of the pocket, while the polar heads are located outside and bind the iron with their C9. From these results, a possible rule as LOX11 could be detected.

For what concerns hypothetical LOX and *Chaetoceros debilis*, docking results highlighted good binding affinity with EPA. The ligand binds the active site with its tail oriented in the deepest portion of the pocket and its polar head outside. The carbon that interacts with the iron is the C7 and the oxidation takes place at the carbon 9, due to the closure of the O2 upper site by the Ala385l; thus, this LOX could be classified as LOX9. Gly537 and Gly556, useful for a deep location of the substrate tail, replace the deepest portion of the pocket flanked by Ala589/620. All the residues involved in the correct position of the ligand are preserved except Ile626 and Leu 627, which are replaced by Thr562/563, probably responsible of the different position of the iron attack despite the one seen for EPA in its interaction with other LOX as the LOX from *P. arenysensis*.

4. Discussion

Putative LOX Sequences Detection, Analysis and Classification

As mentioned at the top of the results section, at the beginning of the work presented, the only sequences annotated as diatoms LOX in databases were the ones from *P. arenysensis*, *F. cylindrus*, and *N. inconspicua*; the latter was wrongly annotated, as demonstrated by our analysis and modelling. However, it is correct to mention that since August 2023, LOX from *Skeletonema marinoi* has been annotated as lipoxygenase too. However, even if at the beginning of the present study this annotation was not available, we still considered LOX from *S. marinoi* in the analysis due to its similarity with the LOX from *P. arenysensis* (35% of sequence identity on 70% of query coverage) and its difference from the LOX of *F. cylindrus*. In fact, it became one of the reference structures chosen for representing its relative diatoms group. Therefore, this novel annotation is a further element that corroborates our sequence detection and increases our confidence about the effective LOX nature of the sequence clustered with it.

For what concerns the LOX sequences from *F. cylindrus* and *P. arenysensis*, it is important to underline that from their analysis, it was possible to notice that although both conserved the LOX domain, these two sequences are very different from each other, as different is their cofactor coordination site. Generally, LOX domains present a coordination site composed of three histidine residues, one asparagine, and one leucine or isoleucine, organized along the sequence in a specific way. In particular, the first two histidine residues are often located in the first half of the LOX domain sequence, spaced by four amino acids, while the second histidine, followed by the asparagine after three residues, is located in the second half of the domain; leucine/isoleucine instead is positioned at the end of the sequence.

The LOX from *F. cylindrus* and the similar sequences (groups 4 and 5) present an insertion between the third histidine and the asparagine, which are spaced out by four amino acidic residues. For what concerns LOX from *F. cylindrus*, the tricky point was represented by the presence of an asparagine residue after the third histidine of coordination that might have suggested a possible role of this residue in the coordination. However, the model of the LOX from *F. cylindrus* created displays that are possibly a rearrangement of the helix forming the active site that hosts these residues in a manner to correctly orient the histidine and the asparagine residues near the iron ion, despite the presence of other residues between them. AlphaFold model (AlphaFold DB code: A0A1E7FK81) for this protein sequence has the disadvantage of presenting the coordination residues in an incorrect orientation, determining the presence of the inserted Ala619 in position for the iron coordination. However, on the other hand, considering the nature of this AI, it could not be otherwise, due to the absence of structures characterized by this insertion in the active site among the ones stored in the PDB database and the absence of the cofactor in the model released. Therefore, it was necessary to exploit other approaches for the building of a model in an active conformation, functional for docking analyses, and that could be a reference for the modelling of similar sequences.

Docking results have shown that different residues in the active site pocket result in different LOX types making different interactions with the substrates. In particular, the similar composition of the active site between LOX from *P. homomalla* and *P. arenysesnis* determines the attack on the same carbon atom (C10) for substrates with the same length. The difference in terms of LOX products is attributable to the different composition of the oxygen channel that allows the oxidation in the deepest portion of the substrate (C12). The result obtained is also in agreement with the literature data, which report a possible role of LOX from *P. arenysensis* as LOX12 or 15, due to the discovery of oxylipins produced by this diatom by stereo- and regio-selective oxidation of EPA at C12 and C15 [47]. An example of this is also given by the hypothetical LOX from *P. tricornutum*, for which, as explained in the result section, the active site presents many differences that can be related to the absence of interaction with EPA and the possible binding with the 9,12-hexa-decadienoic acid. The preference for this substrate, despite the EPA, could be related to the major cellular availability of 9,12-hexa-decadienoic acid by these kinds of diatoms, which produce in principle polyunsaturated fatty acid with 16-carbon atoms [48], and that could have driven the LOX active site for this organism in a composition more suitable for this kind of PUFA.

Similarly, docking analyses highlight that a crucial role in the substrate preference is also played by the composition of the deepest portion of the catalytic pocket. *S. marinoi* LOX, for example, while still preserving the composition of the catalytic area, shows a reduction in the deepest pocket part, that does not allow the substrates to reach deeper into the pocket. This results in a different attack of EPA on the binding side, in a different carbon atom that can interact with the substrate, and in a binding capability, as highlighted by the quite comparable binding energy detected, also for smaller substrates as HTrA. These results are in line with a literature study reporting the possible presence of an LOX9 pathway acting on HtrA in *Skeletonema marinoi* [49].

The impact of the pocket shape and composition on the method of binding the substrates is even more visible in that LOXs that present an insertion among their coordination residues. The slight change in the active site shape, due to the presence of another amino acidic residue, and the difference in terms of residues forming the whole catalytic pocket seem to lead to a different adaptability to different substrates. In LOX from *F. cylindrus*, for example, as demonstrated by the comparable binding energies between the two poses, these changes lead to a different binding conformation of EPA that could result in the generation of different products. In a similar way, LOX from *A. glacialis* seems to be adaptable to different kinds of substrates, even if the lowest binding energy values and the Ki seem to decrease as the length of the substrates increases. The enzyme seems to prefer bigger substrates; however, the difference seems to be imputable not to the pocket depth that results still comparable in this area to the ones detected for LOX from *P. arenysensis* and *S. marinoi*, but to the different compositions of the residues located in the middle of the cavity responsible for the substrate conformation. The pocket shape conformational change seems to also affect LOX from *Bellerochea* sp. that, because of a cavity that is tighter in the central part but wider in the deepest portion, seems to bind bigger ligands better than the smaller ones. LOX from *C. debilis*, instead, despite the insertion, is the one that seems to have the most borderline behavior; the helix break is smaller and there is an inferior conformational change in the pocket that remains in shape quite similar to the one of LOX from *P. arenysensis*, for example. In this case, what changes are some residues in the pocket that again may become crucial for the substrate to bind the pocket. Despite the good binding affinity detected for EPA, consistent with the literature studies [50] that highlighted these diatoms as a considerable producer of this fatty acid, it is probable that products generated from these substrates are typical of a LOX9.

From our results noteworthy are also the LOX type detected for LOX from *P. dubia*. This enzyme seems to reveal a possible preference for bigger substrates such as the DHA and the octacosa-octaenoic acid, a substrate for which this LOX seems to assume the role of a LOX 7 or LOX16, respectively. LOX 7 and LOX16 appear rare but not absent in nature; for example, LOX7 is present in *Z. mays*, while LOX16 has been recently discovered in *Cucumis sativus* L. [51].

5. Conclusions

Diatom lipoxygenase evolution reflects the dynamic interplay between these microorganisms and their environment. Over millions of years, diatoms have fine-tuned their lipoxygenase enzymes to cope with diverse ecological challenges. Evolutionary pressures have driven the development of specific lipoxygenase isoforms tailored to respond to environmental cues, such as nutrient availability and predation threats. The evolution of diatom lipoxygenase has played a pivotal role in shaping the algae's resilience, adaptability, and ecological success. Understanding the molecular evolution of these enzymes provides insights into the intricate mechanisms that allow diatoms to thrive in a wide range of aquatic habitats, contributing to the broader understanding of evolutionary processes in microbial ecosystems. In this context, the description of molecular aspects on the substrate selection may open the way to further experimental analyses for the selection of oxylipins of biotechnological interest.

Supplementary Materials: The following supporting information can be downloaded at: https://www.mdpi.com/article/10.3390/biom14030276/s1, Figure S1: Visualization of the "U-shaped" pocket of the LOX from *P. homomalla* and details of the interactions between substrate and cofactors; Figure S2: Motif research along the sequences belonging to groups 1 and 2.; Figure S3: 3D structural conformation of the cofactor coordination residues present in the structure of LOXs; Table S1: Grid parameters for docking simulations; Table S2: List of the 45 sequences found to be similar to LOXs from *P. arenysensis* or *F. Cylindrus*.

Author Contributions: Conceptualization, D.G., D.D. and A.F.; methodology, D.G. and A.F.; validation, D.G. and A.F.; investigation, D.G., S.B. and I.D.; writing—original draft preparation, D.G.,

D.D. and A.F.; writing—review and editing, D.G., D.D. and A.F.; visualization, D.G., S.B. and I.D.; supervision, A.F. All authors have read and agreed to the published version of the manuscript.

Funding: This research was partially funded under the National Recovery and Resilience Plan (NRRP), Mission 4 Component 2 Investment 1.4—Call for tender No. 3138 of 16 December 2021, rectified by Decree n.3175 of 18 December 2021 of Italian Ministry of University and Research funded by the European Union—NextGenerationEU; Award Number: Project code CN_00000033, Concession Decree No. 1034 of 17 June 2022 adopted by the Italian Ministry of University and Research, CUP C63C22000520001, Project title "National Biodiversity Future Center—NBFC".

Institutional Review Board Statement: Not applicable.

Informed Consent Statement: Not applicable.

Data Availability Statement: The data used in this study were available in resources available in the public domain (listed under the Section 2). The original contributions presented in the study are included in the article/Supplementary Material, further inquiries can be directed to the corresponding author.

Acknowledgments: D.G. was supported within the framework of "CIR01_00017—"CNRBiOmics—Centro Nazionale di Ricerca in Bioinformatica per le Scienze Omiche"—Rafforzamento del capitale umano"—CUP B56J20000960001 and by CNR project NUTRAGE FOE-2021 DBA.AD005.225.

Conflicts of Interest: The authors declare no conflicts of interest.

References

1. Sardo, A.; Orefice, I.; Balzano, S.; Barra, L.; Romano, G. Mini-Review: Potential of Diatom-Derived Silica for Biomedical Applications. *Appl. Sci.* **2021**, *11*, 4533. [CrossRef]
2. Mallimadugula, B.V.; Hameed, A. Diatom Algae for Carbon Sequestration in Oceans. In: Insights into the World of Diatoms: From Essentials to Applications. In *Plant Life and Environment Dynamics*; Srivastava, P., Khan, A.S., Verma, J., Dhyani, S., Eds.; Springer: Singapore, 2023. [CrossRef]
3. D'Ippolito, G.; Nuzzo, G.; Sardo, A.; Manzo, E.; Gallo, C.; Fontana, A. Lipoxygenases and Lipoxygenase Products in Marine Diatoms. *Methods Enzymol.* **2018**, *605*, 69–100. [CrossRef] [PubMed]
4. d'Ippolito, G.; Tucci, S.; Cutignano, A.; Romano, G.; Cimino, G.; Miralto, A.; Fontana, A. The role of complex lipids in the synthesis of bioactive aldehydes of the marine diatom *Skeletonema costatum*. *Biochim. Biophys. Acta* **2004**, *1686*, 100–107. [CrossRef] [PubMed]
5. Cutignano, A.; d'Ippolito, G.; Romano, G.; Lamari, N.; Cimino, G.; Febbraio, F.; Fontana, A. Chloroplastic glycolipids fuel aldehyde biosynthesis in the marine diatom *Thalassiosira rotula*. *ChemBioChem* **2006**, *7*, 450–456. [CrossRef]
6. Fontana, A.; d'Ippolito, G.; Cutignano, A.; Romano, G.; Lamari, N.; Massa Gallucci, A.; Cimino, G.; Miralto, A.; Ianora, A. LOX-induced lipid peroxidation mechanism responsible for the detrimental effect of marine diatoms on zooplankton grazers. *ChemBioChem* **2007**, *8*, 1810–1818. [CrossRef] [PubMed]
7. d'Ippolito, G.; Lamari, N.; Montresor, M.; Romano, G.; Cutignano, A.; Gerecht, A.; Cimino, G.; Fontana, A. 15S-lipoxygenase metabolism in the marine diatom *Pseudo-nitzschia delicatissima*. *New Phytol.* **2009**, *183*, 1064–1071. [CrossRef] [PubMed]
8. Nanjappa, D.; d'Ippolito, G.; Gallo, C.; Zingone, A.; Fontana, A. Oxylipin diversity in the diatom family Leptocylindraceae reveals DHA derivatives in marine diatoms. *Mar. Drugs* **2014**, *12*, 368–384. [CrossRef]
9. Ianora, A.; Miralto, A.; Romano, G. Antipredatory defensive role of planktonic marine natural products. In *Handbook of Marine Natural Products*; Fattorusso, E., Gerwick, W.H., Taglialatela-Scafati, O., Eds.; Springer: Dordrecht, The Netherlands, 2012; pp. 711–748.
10. Russo, E.; Ianora, A.; Carotenuto, Y. Re-shaping marine plankton communities: Effects of diatom oxylipins on copepods and beyond. *Mar. Biol.* **2019**, *166*, 9. [CrossRef]
11. Russo, E.; d'Ippolito, G.; Fontana, A.; Sarno, D.; D'Alelio, D.; Busseni, G.; Ianora, A.; von Elert, E.; Carotenuto, Y. Density-dependent oxylipin production in natural diatom communities: Possible implications for plankton dynamics. *ISME J.* **2020**, *14*, 164–177. [CrossRef]
12. Kâ, S.; Carotenuto, Y.; Romano, G.; Hwang, J.S.; Buttino, I.; Ianora, A. Impact of the diatom-derived polyunsaturated aldehyde 2-trans,4-trans decadienal on the feeding, survivorship and reproductive success of the calanoid copepod Temora stylifera. *Mar. Environ. Res.* **2014**, *93*, 31–37. [CrossRef]
13. Carotenuto, Y.; Ianora, A.; Miralto, A. Maternal and neonate diatom diets impair development and sex differentiation in the copepod *Temora stylifera*. *J. Exp. Mar. Biol. Ecol.* **2011**, *396*, 99–107. [CrossRef]
14. Ianora, A.; Bastianini, M.; Carotenuto, Y.; Casotti, R.; Roncalli, V.; Miralto, A.; Romano, G.; Gerecht, A.; Fontana, A.; Turner, J.T. Non-volatile oxylipins can render some diatom blooms more toxic for copepod reproduction. *Harmful Algae* **2015**, *44*, 1–7. [CrossRef]

15. Asai, S.; Sanges, R.; Lauritano, C.; Lindeque, P.K.; Esposito, F.; Ianora, A.; Carotenuto, Y. De Novo Transcriptome Assembly and Gene Expression Profiling of the Copepod Calanus helgolandicus Feeding on the PUA-Producing Diatom Skeletonema marinoi. *Mar. Drugs* **2020**, *18*, 392. [CrossRef] [PubMed]
16. Russo, E.; Lauritano, C.; d'Ippolito, G.; Fontana, A.; Sarno, D.; von Elert, E.; Ianora, A.; Carotenuto, Y. RNA-Seq and differential gene expression analysis in Temora stylifera copepod females with contrasting non-feeding nauplii survival rates: An environmental transcriptomics study. *BMC Genom.* **2020**, *21*, 693. [CrossRef] [PubMed]
17. Palomera, I.; Olivar, M.P.; Salat, J.; Sabatés, A.; Coll, M.; García, A.; Morales-Nin, B. Small pelagic fish in the NW Mediterranean sea: An ecological review. *Prog. Oceanogr.* **2007**, *74*, 377–396. [CrossRef]
18. Ruocco, N.; Albarano, L.; Esposito, R.; Zupo, V.; Costantini, M.; Ianora, A. Multiple Roles of Diatom-Derived Oxylipins within Marine Environments and Their Potential Biotechnological Applications. *Mar. Drugs* **2020**, *18*, 342. [CrossRef]
19. Linares-Maurizi, A.; Reversat, G.; Awad, R.; Bultel-Poncé, V.; Oger, C.; Galano, J.-M.; Balas, L.; Durbec, A.; Bertrand-Michel, J.; Durand, T.; et al. Bioactive Oxylipins Profile in Marine Microalgae. *Mar. Drugs* **2023**, *21*, 136. [CrossRef]
20. Shipelin, V.A.; Sidorova, Y.S. Oxylipins—Biologically active substances of food. *Vopr. Pitan.* **2020**, *89*, 6–13. [CrossRef]
21. Shearer, G.C.; Walker, R. An overview of the biologic effects of omega-6 oxylipins in humans. *Prostaglandins Leukot. Essent. Fat. Acids* **2018**, *137*, 26–38. [CrossRef]
22. Miralto, A.; Barone, G.; Romano, G.; Poulet, S.A.; Ianora, A.; Russo, G.L.; Buttino, I.; Mazzarella, G.; Laabir, M.; Cabrini, M.; et al. The insidious effect of diatoms on copepod reproduction. *Nature* **1999**, *402*, 173–176. [CrossRef]
23. Sansone, C.; Braca, A.; Ercolesi, E.; Romano, G.; Palumbo, A.; Casotti, R.; Francone, M.; Ianora, A. Diatom-derived polyunsaturated aldehydes activate cell death in human cancer cell lines but not normal cells. *PLoS ONE* **2014**, *9*, e101220. [CrossRef] [PubMed]
24. Newcomer, M.E.; Brash, A.R. The structural basis for specificity in lipoxygenase catalysis. *Protein Sci.* **2015**, *24*, 298–309. [CrossRef] [PubMed]
25. Walther, M.; Ivanov, I.; Myagkova, G.; Kuhn, H. Alterations of lipoxygenase specificity by targeted substrate modification and site-directed mutagenesis. *Chem. Biol.* **2001**, *8*, 779–790. [CrossRef] [PubMed]
26. Maeda, Y.; Tanaka, T. Molecular Insights into Lipoxygenases in Diatoms Based on Structure Prediction: A Pioneering Study on Lipoxygenases Found in *Pseudo-nitzschia arenysensis* and *Fragilariopsis cylindrus*. *Mar. Biotechnol.* **2022**, *24*, 468–479. [CrossRef] [PubMed]
27. Neau, D.B.; Bender, G.; Boeglin, W.E.; Bartlett, S.G.; Brash, A.R.; Newcomer, M.E. Crystal structure of a lipoxygenase in complex with substrate: The arachidonic acid-binding site of 8R-lipoxygenase. *J. Biol. Chem.* **2014**, *289*, 31905–31913. [CrossRef] [PubMed]
28. Johnson, M.; Zaretskaya, I.; Raytselis, Y.; Merezhuk, Y.; McGinnis, S.; Madden, T.L. NCBI BLAST: A better web interface. *Nucleic Acids Res.* **2008**, *36*, W5–W9. [CrossRef] [PubMed]
29. Altschul, S.; Boguski, M.; Gish, W.; Wootton, J.C. Issues in searching molecular sequence databases. *Nat. Genet.* **1994**, *6*, 119–129. [CrossRef]
30. Gasteiger, E.; Gattiker, A.; Hoogland, C.; Ivanyi, I.; Appel, R.D.; Bairoch, A. ExPASy: The proteomics server for in-depth protein knowledge and analysis. *Nucleic Acids Res.* **2003**, *31*, 3784–3788. [CrossRef]
31. Giordano, D.; Facchiano, A. Classification of microbial transglutaminases by evaluation of evolution trees, sequence motifs, secondary structure topology and conservation of potential catalytic residues. *Biochem. Biophys. Res. Commun.* **2019**, *509*, 506–513. [CrossRef]
32. Sievers, F.; Wilm, A.; Dineen, D.; Gibson, T.J.; Karplus, K.; Li, W.; Lopez, R.; McWilliam, H.; Remmert, M.; Söding, J.; et al. Fast, scalable generation of high-quality protein multiple sequence alignments using Clustal Omega. *Mol. Syst. Biol.* **2011**, *7*, 539. [CrossRef]
33. De Castro, E.; Sigrist, C.J.A.; Gattiker, A.; Bulliard, V.; Langendijk-Genevaux, P.S.; Gasteiger, E.; Bairoch, A.; Hulo, N. ScanProsite: Detection of PROSITE signature matches and ProRule-associated functional and structural residues in proteins. *Nucleic Acids Res.* **2006**, *34*, W362–W365. [CrossRef] [PubMed]
34. Jones, P.; Binns, D.; Chang, H.Y.; Fraser, M.; Li, W.; McAnulla, C.; McWilliam, H.; Maslen, J.; Mitchell, A.; Nuka, G.; et al. InterProScan 5: Genome-scale protein function classification. *Bioinformatics* **2014**, *30*, 1236–1240. [CrossRef] [PubMed]
35. Tamura, K.; Peterson, D.; Peterson, N.; Stecher, G.; Nei, M.; Kumar, S. MEGA5: Molecular Evolutionary Genetic Analysis using Maximum Likelihood, Evolutionary Distance, and Maximum Parsimony Methods. *Mol. Biol. Evol.* **2011**, *28*, 2731–2739. [CrossRef] [PubMed]
36. Bailey, T.L.; Boden, M.; Buske, F.A.; Frith, M.; Grant, C.E.; Clementi, L.; Ren, J.; Li, W.W.; Noble, W.S. MEME SUITE: Tools for motif discovery and searching. *Nucleic Acids Res.* **2009**, *37*, W202–W208. [CrossRef] [PubMed]
37. Giordano, D.; Facchiano, A.; Minasi, P.; D'Agostino, N.; Parisi, M.; Carbone, V. Phenolic Compounds and Capsaicinoids in Three *Capsicum annuum* Varieties: From Analytical Characterization to In Silico Hypotheses on Biological Activity. *Molecules* **2023**, *28*, 6772. [CrossRef] [PubMed]
38. Waterhouse, A.; Bertoni, M.; Bienert, S.; Studer, G.; Tauriello, G.; Gumienny, R.; Heer, F.T.; de Beer, T.A.P.; Rempfer, C.; Bordoli, L.; et al. SWISS-MODEL: Homology modelling of protein structures and complexes. *Nucleic Acids Res.* **2018**, *46*, W296–W303. [CrossRef]
39. Eswar, N.; Eramian, D.; Webb, B.; Shen, M.Y.; Sali, A. Protein Structure Modeling with MODELLER. In *Structural Proteomics. Methods in Molecular Biology™*; Kobe, B., Guss, M., Huber, T., Eds.; Humana Press: Totowa, NJ, USA, 2008; Volume 426. [CrossRef]

40. Wiederstein, M.; Sippl, M.J. ProSA-web: Interactive web service for the recognition of errors in three-dimensional structures of proteins. *Nucleic Acids Res.* **2007**, *35*, W407–W410. [CrossRef]
41. Benkert, P.; Tosatto, S.C.E.; Schomburg, D. QMEAN: A comprehensive scoring function for model quality assessment. *Proteins Struct. Funct. Bioinform.* **2008**, *71*, 261–277. [CrossRef]
42. Laskowski, R.A.; MacArthur, M.W.; Moss, D.S.; Thornton, J.M. PROCHECK—A program to check the stereochemical quality of protein structures. *J. App. Cryst.* **1993**, *26*, 283–291. [CrossRef]
43. Mirdita, M.; Schütze, K.; Moriwaki, Y.; Heo, L.; Ovchinnikov, S.; Steinegger, M. ColabFold: Making protein folding accessible to all. *Nat. Methods* **2022**, *19*, 679–682. [CrossRef]
44. Morris, G.M.; Huey, R.; Lindstrom, W.; Sanner, M.F.; Belew, R.K.; Goodsell, D.S.; Olson, A.J. Autodock4 and AutoDockTools4: Automated docking with selective receptor flexiblity. *J. Comput. Chem.* **2009**, *16*, 2785–2791. [CrossRef]
45. Xu, S.; Mueser, T.C.; Marnett, L.J.; Funk, M.O., Jr. Crystal structure of 12-lipoxygenase catalytic-domain-inhibitor complex identifies a substrate-binding channel for catalysis. *Structure* **2012**, *20*, 1490–1497. [CrossRef] [PubMed]
46. Coffa, G.; Brash, A.R. A single active site residue directs oxygenation stereospecificity in lipoxygenases: Stereocontrol is linked to the position of oxygenation. *Proc. Natl. Acad. Sci. USA* **2004**, *101*, 15579–15584. [CrossRef] [PubMed]
47. Sabatino, V.; Orefice, I.; Marotta, P.; Ambrosino, L.; Chiusano, M.L.; d'Ippolito, G.; Romano, G.; Fontana, A.; Ferrante, M.I. Silencing of a Pseudo-nitzschia arenysensis lipoxygenase transcript leads to reduced oxylipin production and impaired growth. *New Phytol.* **2022**, *233*, 809–822. [CrossRef] [PubMed]
48. Batna, A.; Scheinkönig, J.; Spiteller, G. The occurrence of furan fatty acids in *Isochrysis* sp. and *Phaeodactylum tricornutum*. *Biochim. Et Biophys. Acta* **1993**, *1166*, 171–176. [CrossRef] [PubMed]
49. Ruocco, N.; Nuzzo, G.; d'Ippolito, G.; Manzo, E.; Sardo, A.; Ianora, A.; Romano, G.; Iuliano, A.; Zupo, V.; Costantini, M.; et al. Lipoxygenase Pathways in Diatoms: Occurrence and Correlation with Grazer Toxicity in Four Benthic Species. *Mar. Drugs* **2020**, *18*, 66. [CrossRef]
50. Tachihana, S.; Nagao, N.; Katayama, T.; Hirahara, M.; Yusoff, F.M.; Banerjee, S.; Shariff, M.; Kurosawa, N.; Toda, T.; Furuya, K. High Productivity of Eicosapentaenoic Acid and Fucoxanthin by a Marine Diatom Chaetoceros gracilis in a Semi-Continuous Culture. *Front. Bioeng. Biotechnol.* **2020**, *11*, 602721. [CrossRef]
51. Gorina, S.S.; Egorova, A.M.; Lantsova, N.V.; Toporkova, Y.Y.; Grechkin, A.N. Discovery of α-Linolenic Acid 16(S)-Lipoxygenase: Cucumber (*Cucumis sativus* L.) Vegetative Lipoxygenase 3. *Int. J. Mol. Sci.* **2023**, *24*, 12977. [CrossRef]

biomolecules

MDPI

Article

Characteristic Binding Landscape of Estrogen Receptor-α36 Protein Enhances Promising Cancer Drug Design

Adeniyi T. Adewumi and Salerwe Mosebi *

Department of Life and Consumer Sciences, University of South Africa, Private Bag X06, Florida 1710, South Africa; adewuat@unisa.ac.za
* Correspondence: mosebs@unisa.ac.za; Tel.: +27-471-2315

Abstract: Breast cancer (BC) remains the most common cancer among women worldwide, and estrogen receptor-α expression is a critical diagnostic factor for BC. Estrogen receptor (ER-α36) is a dominant-negative effector of ER-α66-mediated estrogen-responsive gene pathways. ER-α36 is a novel target that mediates the non-genomic estrogen signaling pathway. However, the crystallized structure of ER-α36 remains unavailable for molecular studies. ER-positive and triple-negative BC tumors aggressively resist the FDA-approved drugs; therefore, highly potent structure-based inhibitors with preeminent benefits over toxicity will preferably replace the current BC treatment. Broussoflanol B (BFB), a *B. papyrifera* bark compound, exhibits potent growth inhibitory activity in ER-negative BC cells by inducing cell cycle arrest. For the first time, we unravel the comparative dynamic events of the enzymes' structures and the binding mechanisms of BFB when bound to the ER-α36 and ER-α66 ligand-binding domain using an all-atom molecular dynamics simulations approach and MM/PBSA-binding-free energy calculations. The dynamic findings have revealed that ER-α36 and ER-α66 LBD undergo timescale "coiling", opening and closing conformations favoring the high-affinity BFB-bound ER-α36 ($\Delta G = -52.57$ kcal/mol) compared to the BFB-bound ER-α66 ($\Delta G = -42.41$ kcal/mol). Moreover, the unbound (1.260 Å) and bound ER-α36 (1.182 Å) exhibit the highest flexibilities and atomistic motions relative to the ER-α66 systems. The RMSF (Å) of the unbound ER-α36 and ER-α66 exhibit lesser stabilities than the BFB-bound systems, resulting in higher structural flexibilities and atomistic motions than the bound variants. These findings present a model that describes the mechanisms by which the BFB compound induces downregulation-accompanied cell cycle arrest at the Gap_0 and Gap_1 phases.

Keywords: Breast cancer; Tumour Threat; Estrogen receptor-α36; Structure-based inhibitors; Broussoflavonol B; All-atoms MD simulations; Binding mechanism

1. Introduction

The death risk of breast cancer (BC) (14%) and prostate cancer (PC) (9%) has remained significantly high, and these cancers are the most prevalent types of cancer among women and men, especially in the Western world [1]. Generally, the disease can be managed by several therapy strategies, including chemotherapy, radiotherapy, targeted therapy, surgery, combined therapy (CT), prevention, and endocrine therapy (ET) [2]. ET is an inflexion point, including selective estrogen receptor modulators (SERMs), selective estrogen receptor downregulators (SERDs), and aromatase inhibitors (AI) [3,4]. Estrogen therapy is one of the most necessary treatments for estrogen receptor (ER)-positive BC. Estrogen signaling is critical to breast cancer initiation and development. However, abnormal levels of human estrogen and androgen hormones cause carcinogenesis of the breast and prostate, respectively [1,5]. About 70% of all types of tumor expression are transcription factors that activate estrogen binding and regulate various gene expressions during proliferation [3]. Human estrogen receptor alpha (hERα or ER-α66) is a variant of ER that is predominantly expressed in BC [3] compared to the Erβ isoform; therefore, it is a critical target in the assessment of prognosis [6] and the design of a BC therapy strategy. Furthermore, ER-α66 cDNA cloning produces a unique spliced ER-α36 protein of 36 kDa, which lacks the transcriptional activation domains (AF-1 and AF-2) of ER-α66 [7]. Nevertheless, ER-α36 possesses conserved DNA-binding (DBD), ligand-binding (LBD), and dimerization domains [7]. In addition, ER-α36 has three myristoylation sites proximal to its N-terminal [6]. Moreover, it is functionally different from ERα-66 and highly associated with the plasma membrane, acting as a negative effector of estrogen-dependent and -independent transactivation of ER-α66 and ER-β [6,8]. ER-α36 is a dominant-negative effector of ER-α66-mediated estrogen-responsive gene pathways and mediates the non-genomic estrogen signaling pathway through EGFR/Src/ERK [9].

Estrogen receptor alpha is characterized by acquired mutations in hormone-receptor-positive (HR+), negative, and metastatic breast cancer (MBC), which makes anticancer drug design challenging [3]. The treatment of endocrine resistance is complicated, causing ET, and even CT, to be ineffective [10]. About 40% of diagnosed HR+ MBC patients treated with ET relapse over time, while treated prostate cancer patients show inconsistencies in ER expression and efficacy of SERMs [11]. ER-α36 dysregulation is hugely associated with various diseases. ER-α36 correlates with a larger tumor size and more distant metastasis and is associated with advanced disease severity and the poor survival of breast cancer patients [12]. Moreover, ER-α36 acquires resistance to ET drugs by activating non-genomic signaling pathways, which provides insights into why resistance characterizes some breast cancers while others are not improved when using an antiestrogen strategy.

However, the molecular mechanisms underlying tamoxifen resistance to inhibiting the function of ER-α36 are still largely unclear. Another problem is that the impact of ER-α36 mutation on the immune modulatory effect of estrogen and the ER-α36 inhibitors is unknown. Computer-based drug design techniques provide a platform that covers the shortcomings of various wet laboratory-based drug design techniques and methods. Interestingly, proteins' geometry and topology parameters, such as cross channels, interior cavities, and surface pockets, are fundamentally critical for the proteins to play their roles. The impacts of dynamics and motions on the structure and enzymatic function of modeled ER-α36 LBD (ligand-binding domain) upon binding fulvestrant (reference inhibitor) and broussoflavonol B (BFB) were investigated using integrated all-atom molecular dynamics simulations. Fulvestrant is an FDA-approved SERD used to treat post-menopause in hormone-receptor-positive advanced (or metastatic) BC, which is also active in patients treated with therapies other than ET [13].

The ICI 182,780 (fulvestrant, Faslodex), a 'pure' antiestrogen, accelerates the degradation and impairs the dimerization and nuclear localization of ER-α66. However, several studies have reported that ICI 182,780 failed to induce the degradation of ER-α36, presumably because ER-α36 has a truncated ligand-binding domain, lacking the last four helices (helix 9–12) of ER-α66, which are essential for protein degradation induced by ICI 182,780. Therefore, the failure of ER-α36 degradation is a highly possible reason for ICI 182,780 resistance.

However, studies have reported processes that drive acquired resistance to fulvestrant therapy [14]. Recent works suggest that ESR1 and PIK3CA mutations and ER/HER2 crosstalk underly the molecular mechanisms for fulvestrant resistance [15,16]. Broussonetia papyrifera (paper mulberry) is a plant that grows naturally in Asia and Pacific countries that contains numerous flavonoid compounds. Among the compounds, broussoflavonol B (BFB; 5,7,3′,4′-tetrahydroxy-3-methoxy-6,8-diprenylflavone) exhibits various activity types against targets. Such protein targets include tyrosinase, aromatase, secretory phospholipase A-2, nitric oxide synthase, and ER-positive breast cancer [17,18]. Broussoflavonol B induces cell cycle arrest at the G_0/G_1 and G_2/M phases via downregulation of c-Myc protein [18]. The compound induces apoptotic cell death, accumulating the annexin-V- and propidium-iodide-positive cells and the cleavage of caspases [17]. The report further showed that the BFB-based treatment decreased the steady-state levels of ER-α36, restricting the growth of the stem-like cells in ER-negative breast cancer MDA-MB-231 [17]. The studies reported the IC_{50} values of broussoflavonol B, which are assay-specific, and that validation with computational results is unreliable [19]. Moreover, the inhibition dynamics of BFB binding to any target have not been studied, especially ER-α36 and ER-α66. Therefore, in this study, we build a 3D structure (homology model) of ER-α36 LBD (ligand-binding domain) in order to gain insights into its dynamics and the motions of the ER-α36 structure and function compared to ER-α66, hypothesizing their identically conserved LBD. The research gap relating to the dynamics of the mechanism, structure, and function of ER-α36 LBD prompted the undertaking of this project, which could aid in designing new efficacious drugs in the context of the human estrogen receptor ER-α36 LBD.

2. Methodology

2.1. Homology Modeling

ER-α36 is the primary study target, and its 3D conformational structure was unavailable. Therefore, the primary structure (amino acid sequence) of ER-α36 was retrieved from UniProt.org [20] and uploaded onto the SwissModel BLAST database (SMD), automatically searching for possible structural templates (Accession ID: P03372-4), which eventually created a homology model. The ER-α36 auto-model criteria by SMD were primarily based on significant correlation with the target (ER-α36) protein, maximum sequence identity (>30%), good sequence similarity, query coverage, and acceptable crystallographic data. The global model quality estimate (GMQE) and quaternary structure quality estimate (QSQE) were also used to evaluate the quality of the model.

Additional validation methods were used to model ER-α36, including the modeler-10.3 add-on UCSF Chimera tool, ConSurf, and AlphaFOLD (FunFOLD) servers [21]. The three (3) selected templates used to model on the UCSF Chimera-Modeler included 1R5K.pdb [22], 2P15.pdb [23], and 6DF6.pdb [24], which were human X-ray-crystallized ER-α66 LBD proteins. A multiple sequence alignment was performed using the EMBL-EBI ClustalW server [25]. ClustalW uses a BLOSUM matrix of 10 for gap opening and a penalties gap of 0.1 extensions for gap extension [26]. The SMD model was superimposed on the models obtained from the other three methods to validate the protein. Further validation was performed by running a 40 ns molecular dynamics simulation for ER-α36 LBD. A plot of Ramachandran bond angles and torsional strain analysis was obtained using Maestro-13.0 software (http://www.schrodinger.com/; 14 October 2023)

2.2. Active Site Identification

Accurately predicting ligand-binding residues from amino acid sequences is crucial for the automated functional annotation of novel proteins. The possible binding sites of ER-α36 LBD were predicted using web servers, including the Computer Atlas of Surface Topography of proteins (CASTp 3.0) [27] and the FunFOLD method incorporated in the IntFOLD integrated protein structure and function prediction server (version 6.0). CASTp 3.0 provides online services for locating, delineating, and measuring protein structures' geometric and topological properties. The CASTp 3.0 server identifies all surface pockets, interior cavities, and cross channels in a protein structure. The method uses the alpha shape method [28] of computational geometry to identify the topographic features and estimate the available entrance openings' volumes, areas, and sizes. The volume, areas, and sizes are analytically computed using Cannolly's molecular surface model [29] and Richards' solvent-accessible surface model [29]. It calculates the secondary structures of proteins using DSSP [30] by obtaining the residue annotations of proteins from the UniProt database and mapping them to the PDB database. FunFOLD is benchmarked against the top servers and the manual prediction groups tested at CASP8 and CASP9 [21]. The FunFOLD method uses an intuitive approach for cluster identification and residue selection to predict the ligand-binding site of a protein. The IntFOLD-incorporated FunFOLD server integrates into existing fold recognition servers, requiring only a 3D model and lists of templates as inputs.

2.3. Molecular Docking Calculations

The docking calculations were performed for broussoflavonol B and fulvestrant (reference inhibitor) (Figure 1). The molecular structure of BFB was downloaded from PubChem, while the structure of fulvestrant (FULV) was drawn in MarvinSketch (http://www.chemaxon.com, accessed on 28 November 2023) and converted to mol2 format. Broussoflavonol B and FULV have been individually assessed in Molegro Molecular Viewer (MMV) software (http://www.molegro.com/, accessed on 28 November 2023) to ensure suitably displayed bond angles and a hybridization state. The inhibitors were minimized and optimized using Avogadro's steepest descent method and the GAFF force field [31]. The AutoDock tool's graphical user interface was used to generate the grid dimension and center of the box for the final chosen binding pocket. The defined ER-α36 grid box (Å) parameters are center ($X = 6.28$, $Y = -0.55$, $Z = -6.32$) and dimensions ($X = 22.39$, $Y = 24.40$, $Z = 19.66$). In the same vein, the grid box at the active site of the ER-α66 protein was defined as center ($X = 13.14$, $Y = 10.73$, $Z = -21.93$) and dimensions ($X = 25.50$, $Y = 22.18$, $Z = 27.14$). The docking calculations were performed with AutoDock Vina using the Lamarckian genetic algorithm [32]. The protein and ligands were prepared using UCSF Chimera software-1.16 by adding hydrogen and Gasteiger charges and assigning the atom types with the AutoDock tool before the calculations were performed.

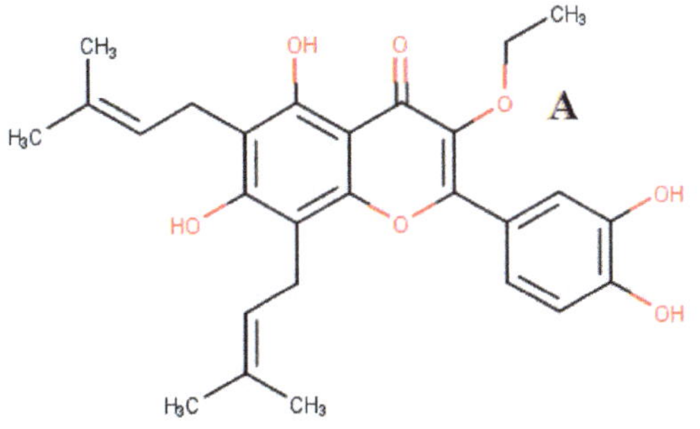

A

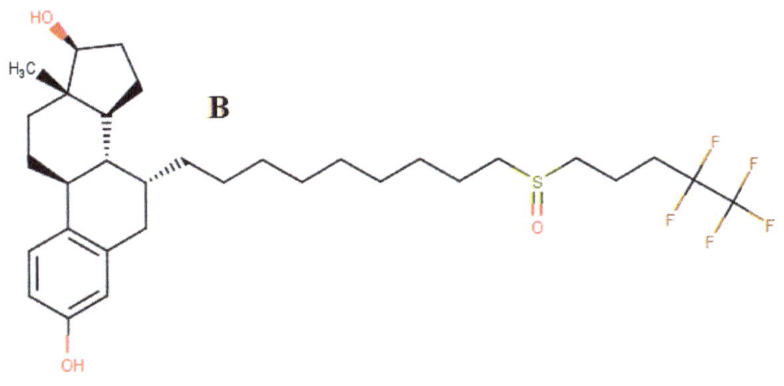

B

Figure 1. Molecular structures of broussoflavonol B (**A**) and fulvestrant (**B**).

2.4. Molecular Dynamics (MD) Simulations [33]

All-atom molecular dynamics simulations were carried out for the ligand-bound BFB-bound ER-α36, BFB-bound ER-α66, and FULV-bound ER-α36 complexes using the best-fit poses docked at their selected active sites. The MD simulations were performed in the AMBER18GPU version of the PMEMD-CUDA engine in the AMBER18 package in the Lengau CHPC server (http://www.ambermd.org, accessed on 28 November 2023) [34]. The hydrogen atoms and Gasteiger charges were added using the UCSF Chimera tool (http://www.cgl.ucsf.edu/chimera, accessed on 28 November 2023). The partial atomic charges for inhibitors BFB and FULV were calculated using the restrained electrostatic potential (RESP), and the ANTECHAMBER module within AMBER was used to generate the force field parameter using the general AMBER force field (GAFF) partial charges [31].

The GAFF assigned atom types and filled in the missing parameters. The molecules and proteins parametrized were the AMBER FF14 force field version [31]. The AMBER LEAP module was utilized to add the protons and the required Na+ ions (counterions) for protein neutralization. Furthermore, the system was explicitly solvated with a box of TIP3P water molecules at a distance of 12.0 Å from the protein atoms [35]. The MD simulations were run using the periodic boundary conditions, while obtaining long-range contributions from the electrostatics from the particle mesh Ewald (PME) method contained in AMBER 18, with direct space and van der Waals interactions restricted to 12.0 Å. The unbound systems and complexes were initially minimized for 2500 steps with 500 kcal/mol $Å^2$ restraint potential, and a whole minimization step of 5000 steps was further run without restraint using the conjugate algorithm. The gradual heating was carried out in a canonical ensemble (NVT) (between 0 and 300 K), executed for 5 ps, fixing several atoms and volumes. A potential harmonic restraint of 10 kcal/mol $Å^2$ was used to restrain the solute systems with a collision frequency of 1.0 ps^{-1}. The systems were equilibrated for 1 ns while keeping the constant operating temperature of 300° K without restraints. All of the hydrogen bonding types were constrained using the SHAKE algorithm. The number of atoms and the pressure were kept constant using an isobaric isothermal ensemble (NPT), Barendsen barostat was used to hold constant pressure at 1 bar, and the MD simulation was carried out over 200 ns [36,37]. The coordinates and trajectories were printed and analyzed every 1ps using the PTRAJ module in AMBER18GPU. The structural conformations of unliganded and complex systems were investigated statistically for stability and flexibility by estimating the root mean square deviations (RMSD) [38] and the root mean square fluctuations (RMSF) [37], respectively. In addition, the atomic displacement of the enzyme was investigated using principal component analysis (PCA). The details of the methods, including the PCA [39], have been reported in our previous works several times. The data were processed with MicroCAL Origin 6.0 software [40], while UCSF Chimera and Discovery Studio 2021 [41] were used for visualization and structural analysis, such as for the ligand-binding mechanism.

2.5. Thermodynamic Parameter Calculations

The binding affinities of the bound and unbound systems were obtained by computing the binding-free energy using the molecular mechanics/GB surface area method (MM/GBSA) [42]. The free energy was calculated based on the average of 10,000 snapshots from a 40 ns trajectory. The estimated binding free energy, ΔG, for each molecular system, including the complex, inhibitor, and protein, can be given as follows [43]:

$$\Delta G_{bind} = G_{complex} - G_{protein} - G_{inhibitor} \tag{1}$$

$$\Delta G_{bind} = E_{gas} + G_{sol} - TS \tag{2}$$

$$E_{gas} = E_{int} + E_{vdw} + E_{ele} \tag{3}$$

$$G_{sol} = G_{GB} + G_{SA} \tag{4}$$

$$G_{SA} = \gamma SASA \tag{5}$$

Moreover, the details of the method can be found in our previous work [38].

3. Results and Discussion

3.1. Structure of the ER-α36 Model and Its Binding Pockets

Some in vitro and in vivo assay findings have been reported on the activity and inhibition of ER-α36, emphasizing the ligand-binding domain (LBD) as a binding pocket. This protein has a significant conserved amino acid sequence with the ER-α66 variant. The amino acids within the LBD include those between 302 and 430. Moreover, the DNA-binding and 27 amino acid domains, whose roles are still unclear, have been reported. Preparing the 3D crystal structure of protein and ligands precedes docking calculations or other steps in ligand-based drug design. Neither the unbound 3D crystal structure of the ER-α36 nor the ER-α36 co-crystallized structure with broussoflavonol B (study inhibitor) was available during this study. Therefore, a homology model was built. The BLASTs of the protein data bank (PDB) database search revealed the 3D templates that we used to build the validated model. A total of 50 templates were used in the SWISS-MODEL algorithm based on the criteria stated in Section 2.1 to build 15 models. The best two models based on satisfactory criteria were selected and superimposed on the Chimera tool for conserved residue comparison of the amino acid sequences. Table 1 details the templates that were used to obtain the best two SWISS-MODELs.

Table 1. Criteria for choosing the templates used to model ER-α36 in each method.

Species	Database ID	Method	Sequence Identity	Sequence Similarity	GMQE	QSQE	Query Cover
			(%)	(%)			(%)
Rattus norvegicus	P06211.1	AlphaFOLD	91.90	0.59	0.80	-	97.0
H. sapiens	PDB7prw	X-ray (2.5 Å)	31.18	0.37	0.59	0.24	89.0
				E-value	Max Score	Total Score	
H. sapiens	1R5K	X-ray (2.7 Å)	89.78	5×10^{-115}	333	333	57.0
H. sapiens	2P15	X-ray (1.94 Å)	89.73	1×10^{-114}	332	332	57.0
H. sapiens	6DF6	X-ray (2.50 Å)	83.66	6×10^{-115}	334	334	62.0
H. sapiens	POS	ConSurf					

Interestingly, the SWISS-MODEL and the ConSurf servers predicted 3D structures without the DNA-binding domain (DBD) and Hinge domain (128 amino acid sequences). In contrast, the AlphaFOLD server and Chimera-Modeler 10.3 add-on model predictions included DBD and Hinge region amino acid sequences. However, deleting the DBD and Hinge domain of the AlphaFOLD server and the Chimera-Modeler models [19] produced the exact same ER-α36 LBD 3D homology model seen in the SWISS-MODEL and ConSurf models.

Regarding the Chimera-Modeler 10.3 methods, three templates were selected to reduce the structural or amino acid differences encountered, using only one template. The regions with the missing residues in the templates were confirmed to be regions outside of the critical active site. A zDOPE value >0.7 reflects ≥95% probability of the model with the correct fold, indicating a reliable model structure, while a negative value indicates a better model. Out of the three models generated on the Chimera-Modeler interface in this study, the best mode gave a zDOPE value of 0.41. A positive zDOPE model might have occurred due to the amino acid sequences of the template. The ConSurf, AlphaFOLD, and UCSF Chimera-Modeler interface models were superimposed with the SWISS-MODEL and observed on the UCSF Chimera tool to look for a significant correlation (Figure 2).

A Ramachandran plot of SWISS-MODEL ER-α36 (Figure 3) was generated in Maestro in order to evaluate the stereochemistry of the ER-α36 structure model. The plot revealed that 97.10% of the residues are located within the most favored region, while 1.74% and 1.16% are located in the additionally and generously allowed regions, respectively. The angle metrics of the ER-α36 residues GLY193 (*phi* (φ) = 93.5° *psi* (ψ) = −12.3°), GLY227 (φ = 81.6°, ψ = −6.1°), and GLY247 (*phi* = 78.2° and *psi* = −3.5°) are found within the generously allowed region. As justified, the angle distribution is good, with more than 95% of the residues occurring within the favored regions, showing suitably modeled ER-α36 geometry and stereochemistry distributions.

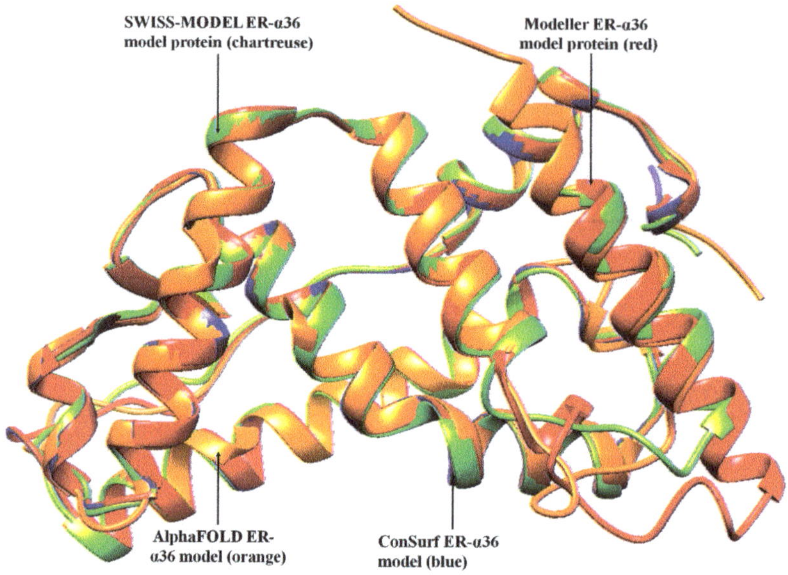

Figure 2. Comparative superimposition of the four ER-α36 models by the SWISS-MODEL, Modeler, AlphaFOLD, and ConSurf databases.

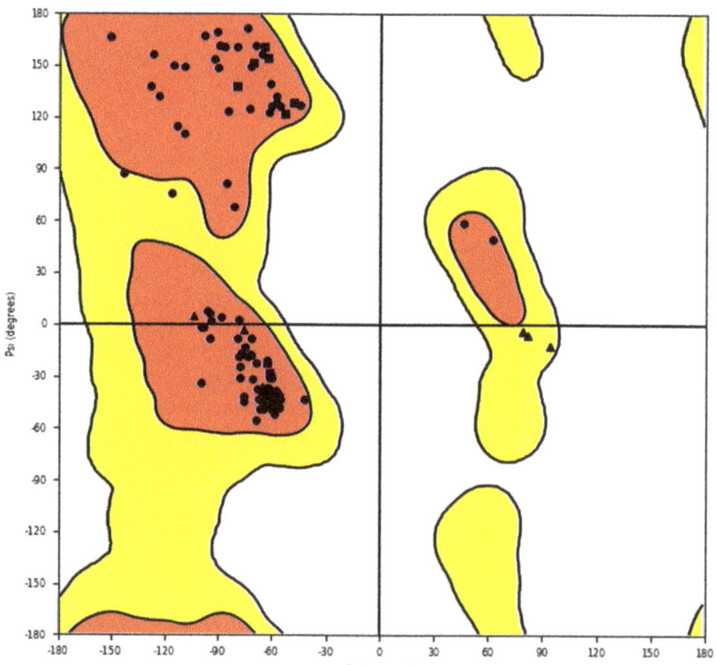

Figure 3. The Ramachandran plot for the homology model of the amino acid sequence of human ER-α36 protein. The red and yellow colours represent the most favoured and allowed regions. The red regions correspond to conformations where there are no steric clashes in the model ER-α36, including the dihedral angles typical of alpha-helical and beta-sheet conformations.

The 3D structures of ER-α36 obtained from the four methods (SWISS-MODEL, Modeler, ConSurf server, and AlphaFOLD) were manually inspected in order to compare their correlation with the 3D structures model from the Psipred outcome (Figure 4). Interestingly, the homology models showed an excellent correlation with the Psipred result.

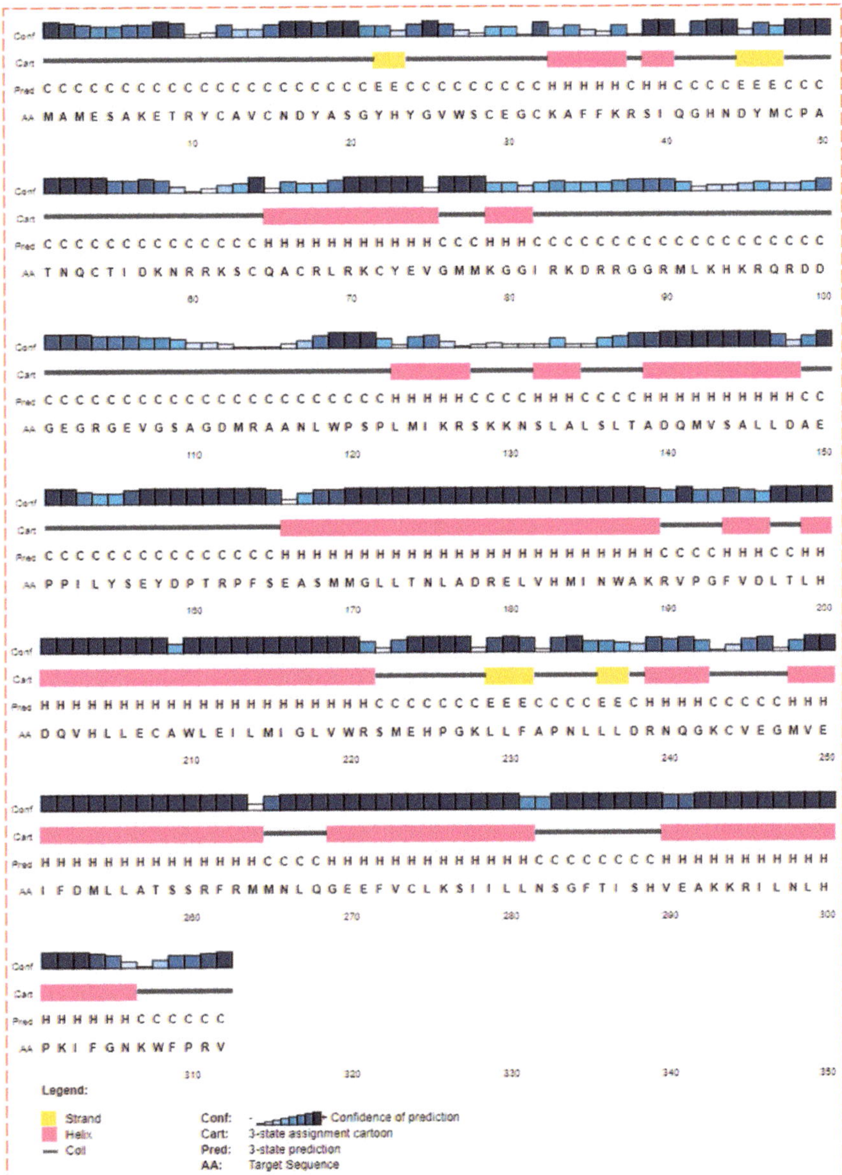

Figure 4. Psipred prediction of the secondary structure of the estrogen receptor-α36.

Excluding the DNA and Hinge domains, the 3D secondary structures of the modeled ER-α36 protein consist of two beta-strands with corresponding residues in the following brackets: β_1-strand (residues: 235–238) and β_2-strand (residues: 227–232). The model has the following eight helices, with the corresponding residues in brackets: α_1-helix (139–149), α_2-

helix (166–190), α_3-helix (194–196), α_4-helix (199–221), α_5-helix (239–244), α_6-helix (248–265), α_7-helix (269–282), and α_8-helix (293–308). Compared to the ER-α66 variant, the ER-α36 LBD lacks helices 9–12; therefore, the ER-α36 protein LBD conformation differs from the ER-α66 [44].

The homology models obtained for the SWISS-MODEL and ConSurf servers did not contain 1–118 residues (residues within the DNA and Hinge domains) [45], yielding an incomplete ER-α36 ligand-binding domain. On the other hand, the AlphaFOLD and Chimera-Modeler models produced full-length ER-α36 protein domains. The ligand-binding domain (LBD) of ER-α36 was fully conserved with ER-α66 LBD, but the truncating of ER-α36 LBD may impact its overall binding of ligands.

Although the pocket residues of ER-α66 have been investigated in order to dock a ligand to a model of ER-α36 LBD based on sequence conservation, the truncated region of the ER-α36 protein confirmed that it has a different conformation. Therefore, the Castp 3.0 server and IntFOLD were used to identify the possible active residues within the ER-α36 LBD protein in order to validate the binding pocket. Both of the active site identification programs predicted almost 100% identical binding residues (Figure 5). The agreement between the Castp and IntFOLD results confirms the locality of the ER-α36 LBD binding pocket and its conserved identity with the ER-α66 LBD. Moreover, the predicted pocket agrees with the reported binding residues of the ER-α36 homology model produced by Kang et al. based on 1ERE.pbd and the templates (ER-α66 LBD) in Table 1, including 7PRW, 1R5K, 2P15, and 6DF6. Figure 6 shows that the amino acids in the ER-α36 LBD pocket are entirely identical to those of the ER-α66 LBD pocket.

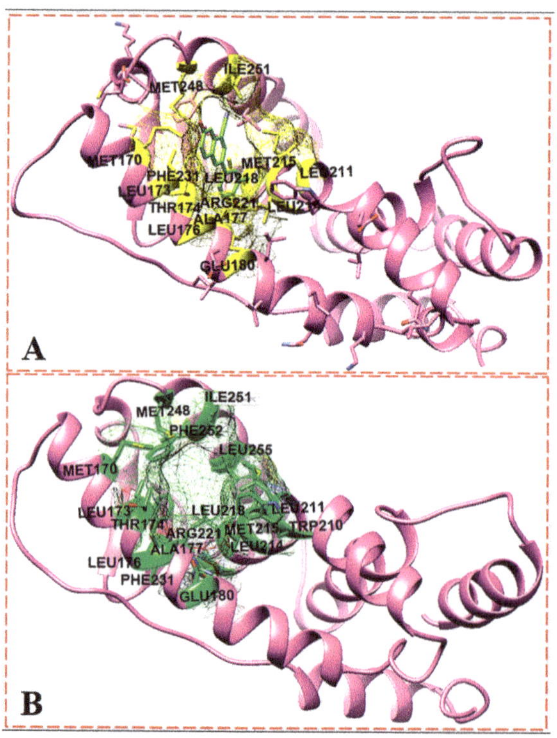

Figure 5. IntFOLD server predicted active site (**A**); Castp server identified active site (**B**); showing the active site residues of ER-α36 protein.

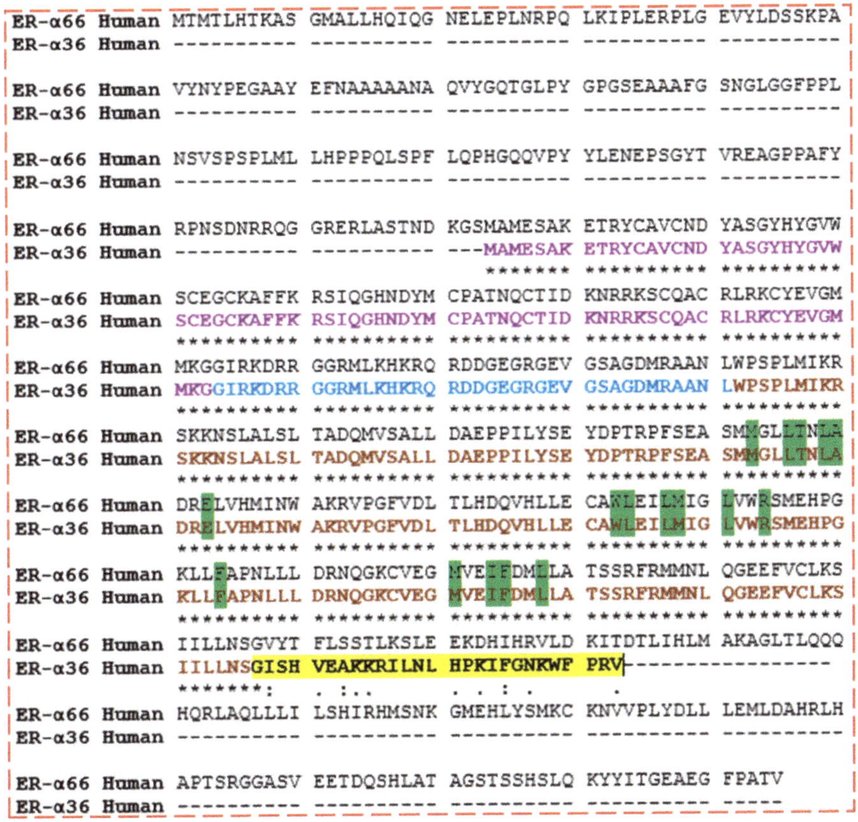

Figure 6. Pairwise sequence alignment of ER-α36 and ER-α66 proteins showing the degree of amino acid conservation, including the unique active sites (green), DNA-binding domain (purple), Hinge domain (blue), ligand-binding domain (LBD) (brown), and 27 unique amino acids (yellow, NOT found in ER-α66 LBD). Fully identical amino acid observed (*), Observed conserved substitutions in the amino acid sequences (:), and Semi-conserved substitutions observed (.).

3.2. Docking Affinity of the Ligands Bound to ER-α36 and ER-α66

Broussoflavonol B and fulvestrant were used in the docking experiments to explore the comparative binding landscape of ER-α36 and ER-α66, even though this study is more critical about the ER-α36 variant. The results have revealed that the poses for both BFB and FULV comfortably fit into the ER-α36 truncated LBD compared to BFB docked to the full-length ER-α66 LBD.

Table 2 shows the binding affinity of broussoflavonol and fulvestrant to their respective ER-α66 and ER-α36 proteins. The results indicate that broussoflavonol had a higher binding affinity (−7.7 kcal/mol) with ER-α36 than ER-α66 and FULV-bound ER-α36. Similarly, it had a higher binding affinity than fulvestrant, with a −7.3 Kcal/mol affinity. The binding affinity values provide insight into the strength of the interaction between the ligands and their respective target molecules, which can contribute to their overall efficacy.

Table 2. The binding affinity of BFB and FULV on their respective enzymes.

Ligand	Target	Docking Scores (kcal/mol)
Brussoflavonol B	ER-α66	−6.5
Brussoflavonol	ER-α36	−7.7
Fulvestrant	ER-α36	−7.3

In the BFB-bound ER-α36 complex, the ligand forms strong polar H-bonds with the side chain of GLU180 and forms hydrophobic interactions, including pi interactions with PHE231, alkyl hydrophobic interactions with ALA177 and LEU211, and several mixed pi/alkyl hydrophobic interactions with TRP210, LEU214, and PHE231. In comparison, the polar residues within the binding pocket of ER-α36 did not show contact with FULV. On the other hand, the BFB-bound complex shows no contact with the ER-α66 LBD polar pocket. These docking data agree with the experimental IC_{50} and reveal the trend in the downregulation of ER-α36 expression inhibition of proliferation in ER-positive breast cancer or growth-inhibitory activity in ER-negative breast cancer [18].

3.3. Molecular Dynamics Simulation Analysis

3.3.1. Thermodynamics Calculations Using MMGB(PB)SA Methods

The experimental IC_{50} data from the in vitro and in vivo study of ER-α36 and the inhibitory activity of the BFB have been reported in various studies, and the binding-free energies of the complex system have never been investigated. In contrast, Kollman et al., 2000, stated that the experimental binding-free energies are more accurate and reliable than the IC_{50} values. Therefore, we computed the binding-free energy using MMGB(PB) methods and compared the correlations with the docking results (Table 2). There was a significant correlation between the MMGB(PB)SA (Table 3) and the docking calculations when comparing the ER-α36 LBD to the ER-α66 LBD for BFB binding. The computed binding-free energies were significantly higher at the BFB-bound ER-α36 LBD site (−52.57 kcal/mol) than those found at the ER-α66 LBD site for BFB (−42.41 kcal/mol). However, the ΔG_{bind} of the ER-α66 LBD for BFB was slightly higher than that found in the ER-α36 LBD for FULV (−37.43 kcal/mol), contrary to the docking results; nevertheless, the difference was much lower when we compared the ΔG_{bind} of the BFB-bound ER-α36 and the BFB-bound ER-α66. Therefore, we conclude that the ER-α36 ligand-binding domain is a novel target, and these interaction mechanisms can be explored to redesign the anticancer FDA-approved drugs facing resistance or mutations or to design newer, promising inhibitors in the clinical stages of drug development.

Table 3. Thermodynamics analysis: summary of the MM/GBSA-based binding-free energy involving ER-α36—BFB, ER-α66—BFB, and ER-α36—FULV complexes.

	Energy Components (kcal mol^{-1})						
Complex	ΔE_{vdW}	ΔE_{elec}	ΔG_{gas}	E_{GB}	E_{SA}	ΔG_{solv}	ΔG_{bind}
ER-α36—BFB	−41.91	38.23	−80.15	34.04	−6.46	27.57	−52.57
	(±4.31)	(±2.02)	(±5.39)	(±3.01)	(±0.43)	(±2.92)	(±4.09)
ER-α36—FULV	−42.64	−12.98	−55.62	24.70	−6.51	18.20	37.43
	(±5.58)	(±7.48)	(±8.59)	(±5.82)	(±0.79)	(±5.64)	(±5.08)
ER-α66—BFB	−48.11	12.65	−60.98	25.28	6.70	18.57	−42.41
	(±5.14)	(±5.25)	(±0.57)	(±5.24)	(±0.57)	(±5.26)	(±4.02)

E_{elec} (electrostatic), ΔE_{vdw} (van der Waals), ΔG_{bind} (calculated total binding free energy), ΔG_{gas} (gas-phase energy), and ΔG_{solv} (solvation-free energy).

3.3.2. Contrasting the Interaction Channels of the ER-α Variants Systems

The 3D interaction channels (Figures 7B, 8B and 9B) indicate the predicted interactions of BFB and FULV with the active sites of the ER-α36 and the interaction networks of BFB with the ER-α66 proteins. The novel ER-α36 truncated binding residues gave significantly more interaction options with BFB than FULV, even compared to the number of interactions between the ER-α66's full-length active site and BFB.

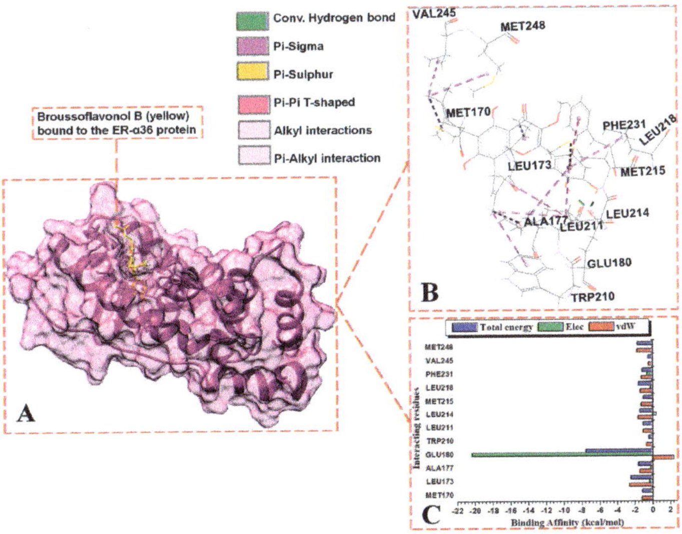

Figure 7. Per-residue energy contributions to the binding of BFB to ER-α36 LBD. (**A**) BFB binds to the active site of the ER-α36, (**B**) interaction network of the BFB-ER-α36 complex, and (**C**) per-residue energy decomposition to BFB.

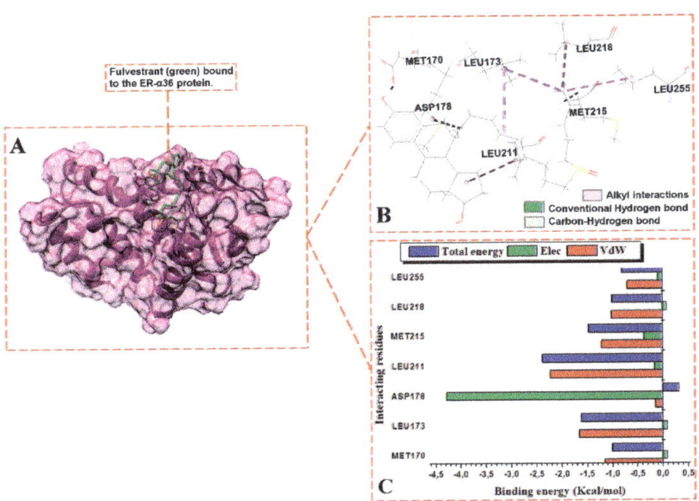

Figure 8. Per-residue energy contributions to the binding of FULV to ER-α36 LBD. (**A**) FULV binds to the active site of the ER-α36, (**B**) interaction network of the FULV-ER-α36 complex, and (**C**) per-residue energy decomposition to FULV.

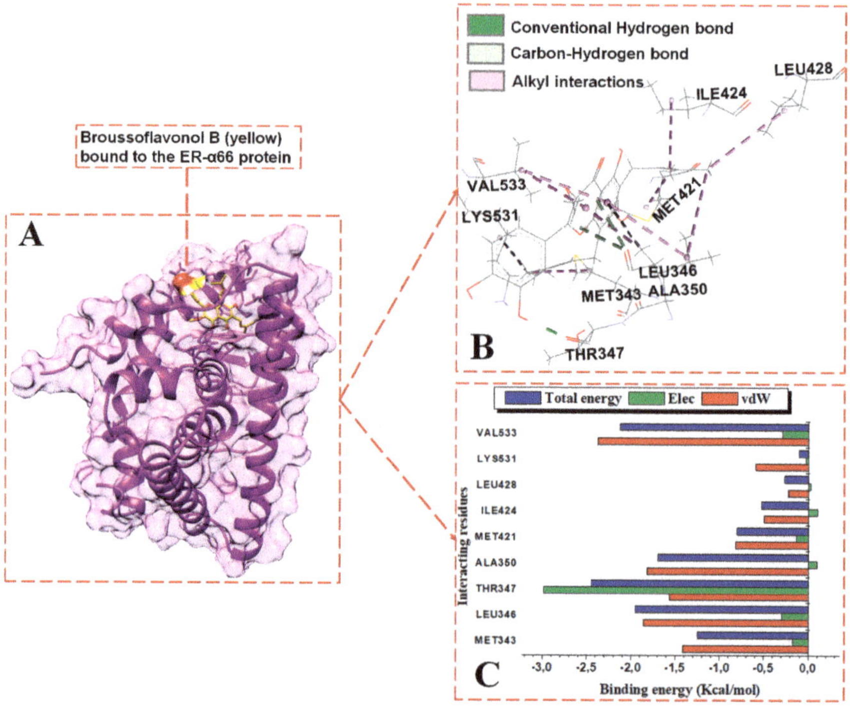

Figure 9. Per-residue energy contributions to the binding of BFB to ER-α66 LBD. (**A**) BFB binds to the active site of the ER-α66, (**B**) interaction network of the BFB-ER-α66 complex, and (**C**) per-residue energy decomposition to BFB.

BFB and FULV interact with ER-α36′s truncated hydrophobic active site, as follows: MET170, LEU173, GLU180, LEU211, MET215, and LEU218. At the binding site of the ER-α36 protein, BFB interacts with ASP178 and LEU255. The following ER-α36 active site residues lie within the ER-α36 novel truncated binding interactions (with their ER-α66 binding pocket equivalent residues enclosed in brackets): MET170 (343), LEU173 (346), THR174 (347), LEU176 (349), ALA177 (350), GLU180 (353), TRP210 (383), LEU211 (384), LEU214 (387), MET215 (388), LEU218 (391), ARG221 (394), PHE231 (404), VAL245 (418), MET248 (421), ILE251 (424), PHE252 (425), and LEU255 (428). The energy contributions in kcal/mol by the ER-α36 pocket residues in binding broussoflavonol B are as follows: MET170 (−1.2), LEU173 (−2.5), ALA177 (−1.63), GLU180 (−7.6), TRP210 (−0.5), LEU211 (−1.12), LEU214 (−1.5), MET215 (−1.05), LEU218 (−1.63), PHE231 (−1.19), VAL245 (−0.53), and MET248 (−1.71). GLU180 contributes significantly higher energy binding for BFB, while TRP210 and VAL245 contribute significantly lower energy binding. The unavailable residues (due to truncation) in the ER-α36 LBD active pocket (which lie within the ER-α66 protein binding site) include GLY521, HIS524, LEU525, ASN532, VAL533, VAL535, PRO535, and LEU539.

The energy contributions in kcal/mol by the ER-α36 pocket residues to fulvestrant binding are as follows: MET170 (−1.0), LEU173 (−1.62), ASP178 (0.31), LEU211 (−2.39), MET215 (−1.48), LEU218 (−1.01), and LEU255 (−0.81). The interactions revealed within the FULV-bound ER-α36 LBD are lower in number and energy values compared to the BFB-bound ER-α36 LBD system. At the full-length-ER-α66-containing binding site, BFB interacts with MET343, LEU346, THR347, ALA350, MET421, ILE424, LEU428, LYS531, and VAL533. The energy contributions in kcal/mol by the ER-α66 pocket residues are as follows: MET343 (−1.25), LEU346 (−2.0.), THR347 (−2.45), ALA350 (−1.7), MET421

(−0.8), ILE424 (−0.5), LEU428 (−0.27), LYS531 (−0.11), and VAL533 (−2.12). Despite the additional interactions from the LYS351 and VAL533 residues with BFB at the ER-α66 binding site, the contributing energies are lower than those from the novel interactions by the binding residues in the BFB-bound ER-α36 complex.

In the BFB-bound ER-α36 complex, the negatively charged GLU180 residue contributes significantly more electrostatic energy (−20.38 kcal/mol) but the lowest total energy compared to the residues in the three complexes. The GLU180 COO^- group interacts with the two hydroxyl oxygen (O) atoms of 2-(3,4-dihydroxyphenyl) substituent adjacent to the 4H-chromen-4-one group of BFB, forming two conventional hydrogen bonds that contribute large electrostatic force and polar solvation energies (Figure 7C). Similarly, the negatively charged ASP178 contributes to the electrostatic and polar solvation energies in the FULV-bound ER-α36 complex by forming an H-bond with an O atom on the FULV phenyl ring. Whereas, in the BFB-bound ER-α66 complex, the THR347 backbone interacts with an O atom on the 2-(3,4-dihydroxyphenyl) group of the BFB to form a conventional H-bond, and the ALA350 backbone interacts with an O atom lying on the 4H-chromen-4-one parent group of the BFB. However, this interaction was not observed with the BF-bound ER-α66 complex. All three of the ligand-bound complexes display hydrophobic interactions. The following residues involve hydrophobic interactions in BFB-bound ER-α36, thus contributing to the overall van der Waals force: MET170 (MET343), LEU173 (LEU346), ALA177 (ALA350), TRP210, LEU211, LEU214, MET215, LEU218, PHE231, VAL245, and MET248 (MET421). The residues in brackets are the ER-α66 residues equivalent to the ER-α36 residues, which form hydrophobic forces with ER-α66. The other residues that form hydrophobic interactions with BFB in the ligand-bound ER-α66 but do not interact in the bound ER-α36 include LEU349, VAL418, MET421, ILE428, LYS531, and VAL533.

For the BFB-bound ER-α36, the sidechain phenyl group of PHE231 forms a pi-hydrophobic (pi-pi T-shaped) interaction with the phenyl substituent adjacent to the parent group of BFB; MET170, LEU211, LEU214, VAL245, and MET248 form alkyl hydrophobic interactions with the 6, 8-bis(3-methylbut-2-en-1-yl) group of the BFB; and LEU173, ALA177, TRP210, and LEU218 form mixed pi/alkyl (pi-sigma and pi-alkyl) hydrophobic interactions. In addition, the sulfur atom of MET215 forms a pi-sulfur interaction with the BFB's 2-(3,4-dihydroxyphenyl) group. MET170, LEU173, LEU211, MET215, LEU218, and LEU255 form alkyl hydrophobic forces with the long aliphatic chains of fulvestrant in the FULV-bound ER-α36 LBD protein. For the BFB-bound ER-α66 LBD system, Met343, MET421, ILE424, LEU428, and LYS531 form alkyl hydrophobic interactions with the 6, 8-bis(3-methylbut-2-en-1-yl) group of the BFB, while LEU346, THR347, ALA350, and VAL533 form pi-alkyl interactions with the adjacent 3,4-hydroxyphenyl group and parent scaffold 4H-chromen-4-one.

3.3.3. Comparative Stability and Flexibility of Estrogen Receptor Variants

In order to further investigate the changes in the secondary structure of the ER-α36 LBD and ER-α66 LBD, we computed the root mean square deviation (RMSD) of the unbound and bound estrogen receptor variants. The RMSD of the C-α atoms of the amino acid backbones is used as an indicator of the stability of a protein structure. The RMSD can estimate the deviations occurring in the backbone atoms of a protein during molecular dynamics simulations. A high RMSD value implies increased atomistic deviations, consequently revealing an unstable protein structure. At the same time, a low RMSD value correlates with a decreased atomistic deviation, implying a stable structural protein system.

Figure 10 shows the binding of the BFB-induced conformational alterations in the ER-α36 LBD and the ER-α66 LBD across the 200-ns MD simulation time and the reference inhibitor (FULV)-induced structural changes in the ER-α36 LBD. The evidence is seen with the conspicuous deviations within the residues' backbone atoms of the BFB-bound ER-α LBD variant proteins compared to its unliganded form. In addition, averaged RMSD (Å) values of 2.147 and 2.024 were computed for the apo ER-α36 LBD and apo ER-α66 LBD, respectively, indicating that the ER-α66 LBD is more compact than the ER-α36 LBD. The BFB-bound ER-α36 and FULV-bound ER-α36 LBD showed higher RMSD (Å) values of

2.089 and 2.273, respectively. At the same time, 2.039 Å was estimated as the RMSD value of the BFB-bound ER-α66 LBD complex. The BFB-bound ER-α66 LBD is a more compact bound enzyme among the bound systems. Although the difference in the RMSD values of the bound and unbound systems is less than 1.0 Å, Figure 11A–D (visual snapshots) indicates significant structural alterations upon the binding of the inhibitors to either of the estrogen receptor variants. The estrogen receptor LBD structures become more stable upon binding to the BFB, unlike when fulvestrant binds to the ER-α36 LBD. However, the ER-α36 LBD was shown to be less stable compared to the ER-α66 LBD.

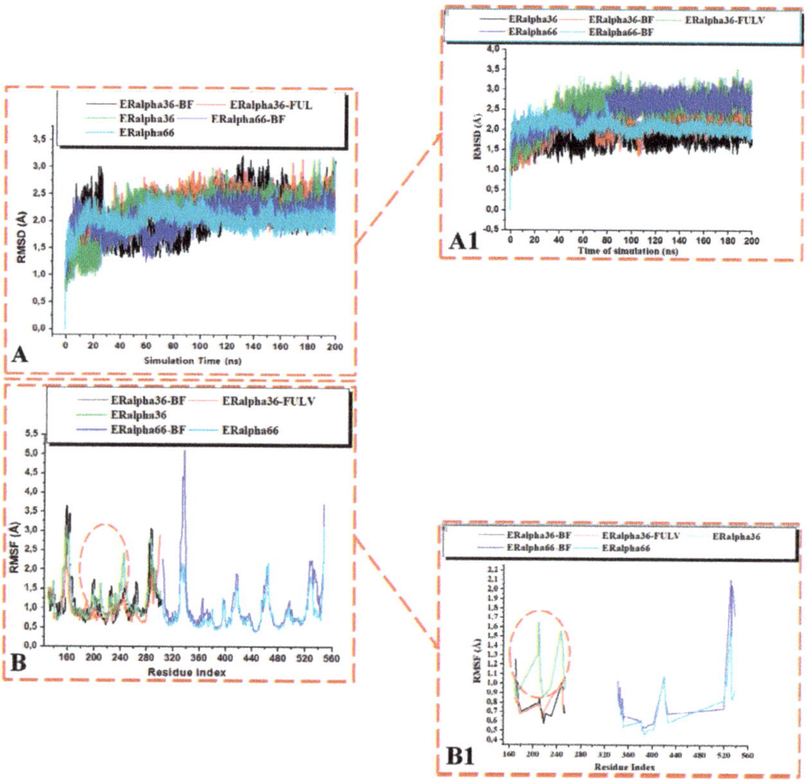

Figure 10. RMSD plots of the whole (**A**) and the active site (**A1**) for the Apo and whole ligand-bound ER-α36 LBD and ER-α66. RMSF plots of the whole (**B**) and the active site (**B1**) for the Apo and ligand-bound ER-α36 LBD and ER-α66. BFB-bound ER-α36 (black), FULV-bound ER-α36 (red), ApoER-α36 LBD (green), BFB-bound ER-α66 (blue), and ApoER-α66 LBD (cyan).

Moreover, the higher stability of both the unliganded and the bound ER-α66 LBD than the ER-α36 LBD could be due to LYS531 and VAL533, which are absent in the ER-α36 partial LBD. These alterations agree with the previous studies and can impact estrogen receptor signaling and expression in breast cancer initiation and development [17,18]. The RMSD trajectories also showed that the ER-α36 partial LBD pocket converged at 20 ns, compared to the ER-α66 LBD, which converged at about 30 ns, and the difference in their fluctuations is no greater than 0.65 Å (Figure 11).

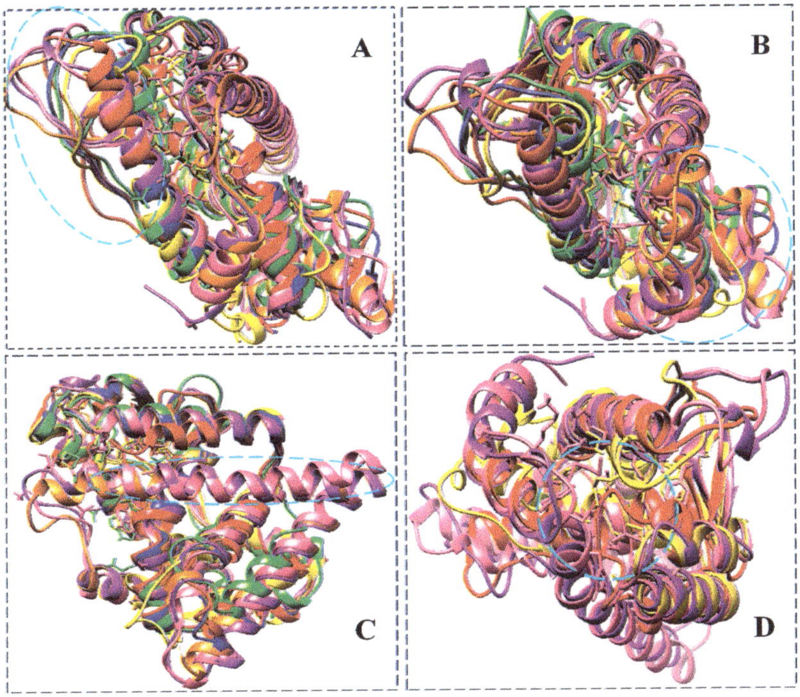

Figure 11. Superimposition of unbound ER-α36 LBD (blue) pre-MD, unbound ER-α36 LBD (red) at 200 ns, BFB-bound ER-α36 LBD (gold) at 200 ns, FULV-bound ER-α36 LBD (lime green) at 200 ns, unbound ER-α66 LBD (orange) pre-MD, ER-α66 (purple) at 200 ns, and BFB-bound ER-α66 LBD (hot pink), showing distinct conformational changes. Insets: active site proximal loop extended to α_2-helix (**A**,**B**). ER-α66 LBD Helix-12 not available in the ER-α36 LBD. The helix-12 of the ER-α66 and (**C**) moved to cover the active site of the ER-α66 (**D**).

Figure 11A–D illustrates the binding poses of BFB and FULV at the active site of the ER-α36 LBD and the ER-α66 LBD compared to the unbound systems at 200 ns. Furthermore, we investigate the fluctuation behavior of the estrogen receptors by estimating the RMSF value, which indicates the mobility of every residue in the protein structure. In comparison to fulvestrant at the ER-α36 LBD site, the RMSF plot (Figure 10B) of broussoflavonol B at both the ER-α36 truncated LBD and the ER-α66 complete LBD site shows a more significant displacement of the residues. The mean RMSF values of the unbound ER-α36 and ER-α66 systems were 1.260 Å and 0.848 Å, respectively. Moreover, the higher RMSF value of the BFB-bound ER-α36 was critically impacted by the residue GLU180 (Figure 10B). More significant movement of all of the residues was seen when BFB bound to the novel truncated ER-α36 LBD (mean RMSF = 1.182 Å) compared to the BFB at the site of the complete ER-α66 LBD (mean RMSF = 1.037 Å). Unlike the ER-α36 site that showed decreased RMSF upon BFB and FULV binding, ER-α66 showed increased fluctuation when BFB bound to its pocket. There is not much flexibility difference in either of the ER-α binding sites, except in the ER-α66 (residues: 330–340 and 528–540) and in the FULV-bound ER-α36 (residues: 190–250), where slightly larger fluctuations occurred.

3.3.4. Analysis of RoG, PCA, and DSSP

To gain further insights into the comparative characteristic 3D conformations of the studied ER-α variants, we computed the radius of gyration (RoG), principal component analysis (PCA), and the defined secondary structure of the protein (DSSP) for the liganded and unliganded systems. The C-α radius of gyration (RoG) measures the structural compactness at the binding interfaces of ER-α36 and ER-α66, in which a less tightly packed protein structure resulted, due to a high RoG value and increased mobility. In contrast, a low ROG value indicates more tightly packed estrogen receptor conformation and reduced mobility.

Figure 12A shows the RoG plots of the ligand-bound ER-α36 systems. The unbound estrogen receptor variants showed diminished RoG values compared to their corresponding bound complexes.

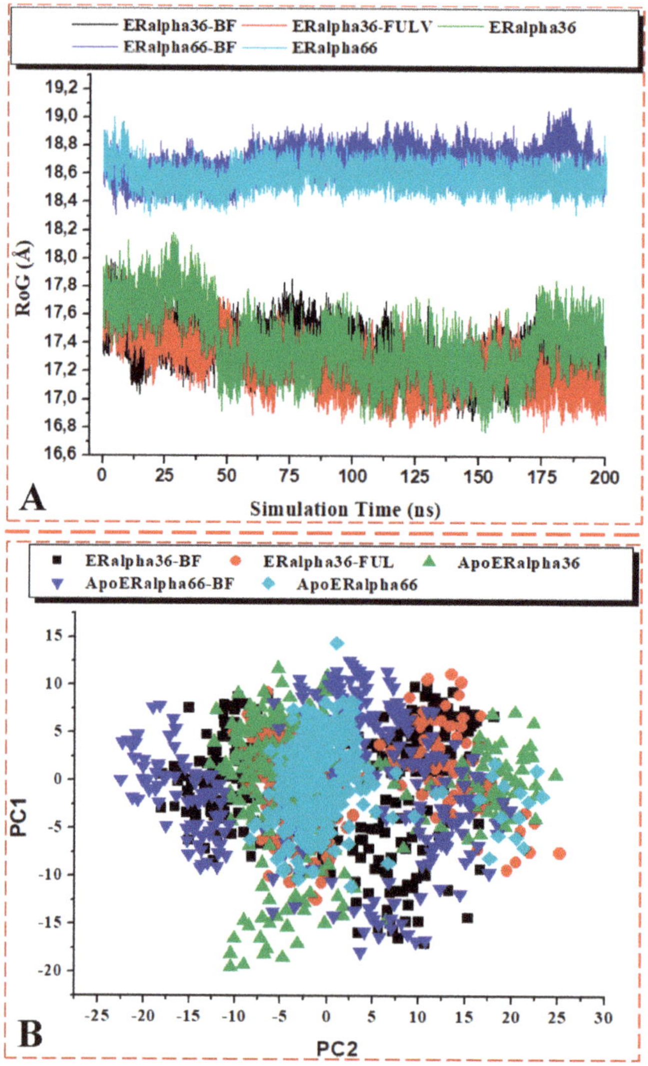

Figure 12. (**A**) C-α atoms' RoG and (**B**) PCA scatter plots of 2000 snapshots along the first two principal components of the ER-α36 LBD and ER-α66 LBD systems. ER-α36–BFB (black), ER-α36–FULV (red), unbound ER-α36 (green), ER-α66–BFB (blue), and bound ER-α66 (cyan). PC1 and PC2, respectively, show differences in motion among the bound and unbound ER-α variants over the simulation time.

The average RoG (Å) of the BFB-bound ER-α36, FULV-bound ER-α36, apo ER-α36, BFB-bound ER-α66, and apo ER-α66 systems were 17.338, 17.239, 17.399, 18.686, and 18.588, respectively. The values imply no significant difference between the bound and unbound respective estrogen receptor variants. However, the bound ER-α36 showed decreased RoG values, while the bound ER-α66 showed increased RoG values when each variant was compared to their respective apo structures. Clearly, the BFB-bound ER-α66 showed a significant RoG difference compared to the BFB-bound ER-α66, which indicates that the ER-α36 protein is a more tightly packed system with reduced mobility. In general, the ER-α36 protein showed a diminished and lower RoG value than ER-α66, indicative of a more tightly packed and less mobile protein. These findings further justify the above-estimated flexibility results from the RMSF analysis, in which a more tightly packed and reduced mobility structure should fluctuate less than a less tightly packed and increased mobility system.

Figure 11A–D elucidates the visual motion shifts across two principal components for the unbound and ligand-bound systems. PCA is a critical analysis for studying simulation trajectories over a time range. The first two principal components (PC1 and PC2) were computed from the 200-ns MD trajectories of the ER-α protein C-α atoms, and the conformational patterns of the bound and unbound receptor variants were projected along the first two eigenvectors (ev1/PC1 versus ev2/PC2) using the C-α atoms' Cartesian coordinates. The eigenvalues represent variance magnitude, and the eigenvectors indicate variance direction. The PCA scatter plots were created using Origin 6.0. The eigenvectors computed from the MD trajectory for all of the systems vary, indicating the protein motion difference between the two ER-α variants. The most significant fluctuation modes with the bound and unbound ER-α protein systems' motion were determined, as shown in Figure 12B.

The conformational behaviors of the bound and unbound ER-α protein variants were projected directionally along PC1 and PC2 to gain insights into the separation of their motions. The PCA plots showed a distinct motion separation in the essential subspace along both principal components among the five ER-α protein systems, with the bound complexes exhibiting highly dispersed motion compared to the unbound proteins. This significant difference is evident from the characteristic structures along the two components. The unbound ER-α36 variant appears to be more dispersed than ER-α66 (more compact motion), concentrated at the center of the two eigenvectors. In straightforward terms, the binding of BFB to either of the ER-α variants showed highly dispersed motions. The induced dispersions in the BFB- and FULV-bound systems could indicate active conformations in the protein, which were not seen in the apo ER-α66 protein, due to compactness and immobility. Inferentially, the active conformation of the protein, such as the whole ER-α and the pocket landscape, could weaken the binding affinity of the ligand.

Therefore, these findings further reflect that the binding of BFB and FULV stabilizes a highly active protein conformation favoring ligand binding, which corroborates the above-explained analytical results. The high ER-α inter-residual displacement induced by the BFB tumor might have resulted in the cell cycle arrest reported in the experimental wet lab studies.

3.3.5. Analysis of DSSP, Distance, and Torsion Angles of the ER-α36 Variant

Furthermore, a comparative defined secondary structure protein (DSSP) analysis was performed on the 200-ns trajectories of the bound and unbound whole ER-α36 and ER-α66 protein systems in order to monitor the conformational changes in their secondary structures. The prior MD and the 200-ns MD structures of the unbound and bound ER-α36 revealed significant structural differences. Figure 13 shows some visual changes in the ER-α36 protein systems, comparing the prior-MD, post-MD, unbound, BFB-bound, and FULV-bound systems. Changes such as twisting, coiling, shifting, and secondary structure transition from one element type to another were observed, particularly in the BFB-bound ER-α36 (green).

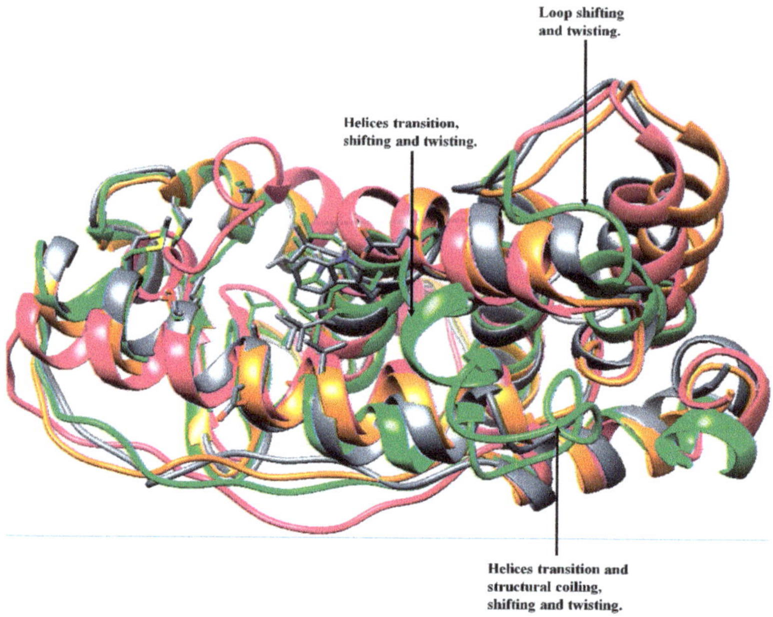

Figure 13. Visual secondary structure changes: superimposed structures of the prior-MD Apo ER-α36 (orange), 200 ns of Apo ER-α36 (deep pink), BFB-bound Apo ER-α36 (green), FULV-bound ApoER-α36 (slate gray), Apo ER-α66 (dark cyan), and BFB-bound ER-α66 (purple) systems.

The visual observations of the loop (proximal to the active site) LYS243, CYS244, VAL245, GLU246, GLY247, MET248, VAL249, GLU250, and ILE251 showed that it shifted and twisted to cover the unbound ER-α36 system at 200 ns (deep pink). Likewise, the BFB-bound ER-α36 loop that contains residues SER283, GLY284, PHE285, THR286, ILE287, SER288, HIE289, VAL290, and GLU291 and the loop that contains residues LYS189, ARG190, VAL191, PRO192, GLY193, PHE194, VAL195, ASP196, LEU197, THR198, LEU199, and HIE200 were visually observed to pull toward the binding pocket in the presence of broussoflavonol B.

Further visual inspections showed that LEU201, LEU202, GLU203, CYS205, ALA206, ILE213, LEU214, MET215, MET248, VAL249, and GLU250 of the unbound ER-α36 transitioned into the loop, while the loop structure of LEU133, ALA134, and LEU135 and SER136 of the apo ER-α36 at 200 ns transitioned into the helices. Due to the interactions of BFB with two helical structures at the active site, these changes caused significant conformational changes in the ER-α36 compared to the remaining systems. In contrast, LEU199, HIS200, VAL290, and GLU291 of the BFB-bound ER-α36 transformed from the helix into the loop.

The binding of a ligand to a protein often induces structural changes to said protein. These conformational changes were further investigated by computing comparative defined secondary structures of the systems (Figure 14A–E). The structural assignment of the systems involves the prominent secondary structure elements, including alpha helices, beta sheets, loops, and coils, which can be sub-categorized as bend, alpha, helices (pi and 3_{10}), turn, anti, para, and none structures. The figures show the positions of each element present in the apo and bound ER-α36 and ER-α66 structures during the 200-ns MD simulation.

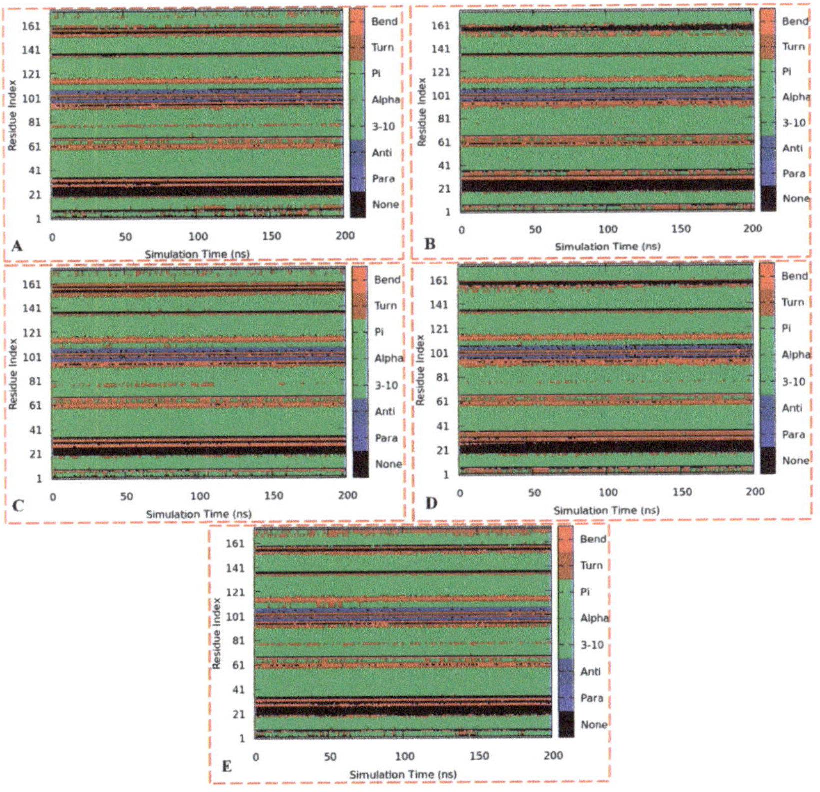

Figure 14. The trajectory DSSP plots of the apo ER-α36 LBD (**A**), apo ER-α66 LBD (**B**), BFB-bound ER-α36 LBD (**C**), BFB-bound ER-α66 LBD (**D**), and FULV-bound ER-α36 LBD (**E**).

Figure 14A,B show the DSSP plots for unbound ER-α36 and ER-α66 systems, while Figure 14C–E represent the bound complexes BFB–ER-α36, BFB–ER-α66, and FULV–ER-α36, respectively. As shown in the plots, there are similarities in the secondary structures of ER-α36 and ER-α66, especially the alpha, para, and bend structures. However, we can observe significant differences, particularly changes in the residues 1–16, 31–36, 76–81, and 155–181. The number of residues transitioning from one secondary structure to another is vital between the unbound proteins and compared with the bound systems. These findings may account for the stability and flexibility discussed in Section 3.3.3, especially comparing the notable changes between the unbound and bound ER-α36 and the conspicuous structural differences between the unbound and bound ER-α66. Therefore, these insights could aid in designing potential small-molecule inhibitors against cancer tumors.

Furthermore, we have comparatively investigated the conformational changes and space of the whole and binding pockets of the truncated ER-α36 LBD and completed ER-α66 LBD snapshots seen in Figure 15A–E. The distance and torsion angle geometric approaches were used to investigate four of the residues of the estrogen receptors that are critically significant to the binding affinity of the BFB and FULV. However, the bond length and torsion angles of the four residues can be extrapolated to the geometry of the remaining residues, because reports have shown that slight variations occur among the residues of biomolecules.

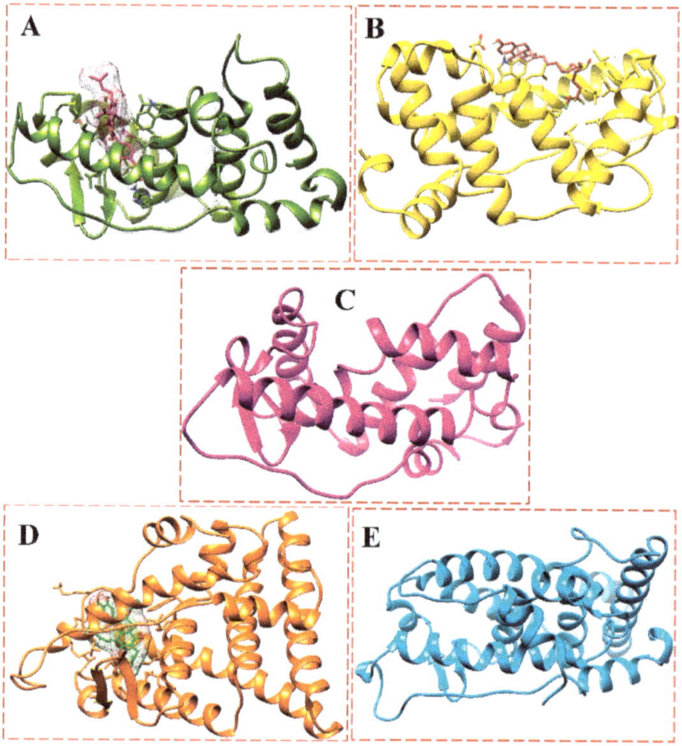

Figure 15. Snapshots at 200-ns MD simulations of apo and complexes: ((**A**), dark green) BFB-bound ER-α36 LBD, ((**B**), gold) FULV-bound ER-α36 LBD, ((**C**), deep pink) unliganded ER-α36 LBD, ((**D**), orange) BFB-bound ER-α66 LBD, and ((**E**), deep sky blue) unliganded ER-α66 LBD.

Protein spontaneously samples the conformational space by variations in torsional angles, significantly impacting the lower energy barrier. Torsion angles are the natural degrees of freedom of protein structures, which provide an understanding of the mechanisms of functional conformational changes or computational models of protein dynamics. Moreover, torsion angles describe the biologically relevant quantities better than atomic coordinates (x, y, z), and the geometric relation of two parts of a molecule can be joined, though it is the most uncomplicated method used to define atoms' position that does not naturally sample the conformational space.

Accordingly, the predicted torsion angles can accelerate protein folding and structure prediction to improve structural precision. However, the ease of obtaining torsional variations that correspond to observed changes in Cartesian coordinates is helpful in protein studies. In addition, the torsion angle of a residue's backbone can impact all of the other protein angles and influence the protein conformation upon ligand binding.

Therefore, the torsion angles of the C-α of LEU173, GLU180, LEU214, and PHE231 were computed in this study. Figure 14 shows the ligands binding to the active sites of the ER-α LBD proteins compared to the unliganded protein structures. The ER-α36 LBD C-α atoms' interacting residues involve hydrogen and oxygen bonding and include LEU173, GLU180, LEU214, and PHE231, with the following corresponding ER-α66 LBD residues: LEU346, GLU353, LEU387, and PHE404, respectively. The 4H-chromen-4-one and 2-(3,4-hydroxyphenyl) of the broussoflavonol remain buried within the hydrophobic active site of the ER-α36 and ER-α66 proteins, while the aliphatic end remains on the surface of the pocket. In contrast, the fulvestrant fluorine atoms protrude into and interact with the hydrophobic pocket of the ER-α36 LBD.

The visual observations of the MD trajectory snapshots show that the conformation of the unbound ER-α LBD active sites is characterized by twisting and the transitional recoiling of the loop covering the binding cavity. Therefore, we computed the distance and angle metrics to accurately describe the motions in order to understand the binding landscape of the estrogen receptors instead of using the distance metrics alone. We have described the dynamics and motion of the comparative binding pockets of the estrogen receptors using the distance and TriC-α angles' (θ, ϕ) combination. Here, Figure 16 illustrates the parameters used to describe the motions, considering the *distance* (d) between the two critical pocket residues (GLU180 and PHE231) relative to *angles* θ (LEU173, GLU180, and LEU214) and ϕ (LEU173, GLU180, LEU214, and PHE231). These computational parameters have been extensively reported as a better option to describe the motion of the protein flaps and dimers in many dimensions, even in one of our previous works.

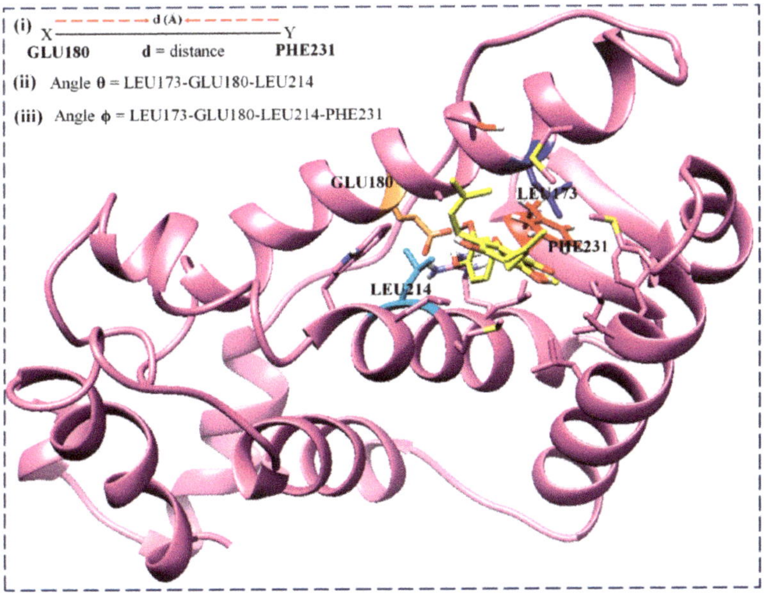

Figure 16. Comparatively defining the binding landscape of the ER-α LBD variants using the distance and angle metrics. Residue color: LEU173 (blue), GLU180 (sandy brown), LEU214 (deep sky blue), and PHE231 (red).

Table 4 confirms the estimated distances between the critical LBD residues GLU180 and PHE231 of the ER-α36 LBD and the estimated distances between their corresponding residues in the ER-α66 LBD. The distances, d_0 (prior to the MD simulations), between the active site residues GLU180 and PHE231 of the unbound enzymes ER-α36 and ER-α66 were 7.743 Å and 7.406 Å, respectively. Both systems had partial openings at 1 ns, but ER-α36 showed a full further pocket opening at 10 ns (d_{10} = 5.882 Å) as it opened very wide (d_{10} = 11.170 Å). However, the ER-α36 active pocket opened again at 50 ns, and the ER-α66 pocket opened at the same simulation time. After another 50-ns MD simulation, we observed that the two pockets closed by shifting to the same positions observed at the first 50 ns. Therefore, both the ER-α36 and the ER-α66 binding activities could be said to undergo opening and closing conformations. However, ER-α36 showed a tighter binding pocket than ER-α66, as evident from the distances at the simulation time events (Table 5).

Table 4. Distance, d, between C-α atoms of GLU180 and PHE231 of the ER-α variant systems.

Simulation	Distance (Angstrom)			
Time (ns)	ER-α36	ER-α36–BFB	ER-α66	ER-α66–BFB
0	7.743	7.743	8.817	7.406
1	7.158	7.728	8.784	7.959
10	5.882	8.352	11.170	10.316
50	8.385	8.187	12.493	12.375
100	5.882	8.352	11.170	10.316
150	5.722	8.520	9.782	11.532
200	8.963	7.701	12.041	11.866

Table 5. Combined TriC-α and dihedral angles relating to simulation time events of the binding cavities of the bound and unbound ER-α66 and ER-α66 enzymes.

Simulation	Bond Angle (θ°)				Torsion Angle (φ°)			
Time (ns)	ER-α36	ER-α36–BFB	ER-α66	ER-α66–BFB	ER-α36	ER-α36–BFB	ER-α66	ER-α66–BFB
0	91.07	65.54	58.65	69.67	−17.33	−22.56	−13.83	−15.57
1	112.46	87.56	82.39	39.63	−25.96	−14.55	−22.86	−32.88
10	53.87	56.86	70.41	89.27	−16.88	−26.99	−30.06	−27.32
50	65.66	54.72	25.98	100.18	−22.55	−30.81	−40.11	−28.56
100	53.87	56.86	70.41	89.27	−16.88	−26.99	−30.06	−27.32
150	75.80	62.57	131.03	103.78	−07.32	−18.45	−40.58	−25.01
200	59.24	60.28	70.93	55.98	−28.21	−16.90	−28.39	−24.22

θ = theta, φ = torsion.

Interestingly, this is the first work that has observed the asymmetric opening of the ER-α36 and ER-α66 active pockets. The asymmetrical opening of the pockets is characterized by various structural transitions and even coiling and twisting, corresponding to a significant shift in the torsional angle of ER-α36 and the ER-α36 at various simulation time ranges. These insights describe the intensive movements of the interacting residues buried within and proximal to the binding cavities, which might create the steric hindrance that causes the opening of the active site.

Furthermore, Table 5 shows the angle θ and dihedral angle ϕ of the bound and unbound ER-α36 and ER-α66 pocket residues, whose impact can be extrapolated to the whole enzyme at a time range of 1–200 ns. It is evident that the distance is inversely proportional to the angle θ metrics but increases as ϕ increases or decreases as ϕ decreases. Overall, there seems to be a large shift in the dihedral angle (ϕ) of the bound and unbound ER-α66. In the bound estrogen receptors, LEU173, GLU180, LEU214, and PHEE231 tightly embraced the ligand BFB without a significant fluctuation from the residues.

Without a doubt, this study will ultimately contribute to designing potent structure-based inhibitors against estrogen receptor variants, especially ER-α36, by targeting the four residues and critical secondary structures, such as the loop covering the cavity. The design will be carried out to inhibit the opening and closing of the active sites.

The computational metrics used in this study provide insights into the inducing effect of BFB by unravelling the molecular mechanism of action of the compound to corroborate the experimental cytotoxic effects (IC_{50}) of BFB at the Gap_0 and Gap_1 phases reported by Jeong and Ryu [18] and Guo et al. [17]. Put together, the significantly higher binding-free energy of BFB upon the binding to ER-α36 (compared to FULV-ER-α36 and BFB-bound ER-α66) and various ER-α36 conformational changes, as given by RMSD, RMSF, RoG, PCA, and DSSP, indicate the impact of BFB on the cell cycle activity. Therefore, these findings provide a model that describes the mechanisms by which the BFB compound induces downregulation-accompanied cell cycle arrest at the Gap_0 and Gap_1 phases when it binds to the ER-α36 protein.

4. Conclusions

The inhibition of negative breast cancers by downregulating the estrogen receptor-α36 ligand-binding domain (LBD) with a few inhibitors (e.g., broussoflavonol B) and FDA-approved drugs (e.g., tamoxifen and fulvestrant) has been reported extensively. Nevertheless, there is no crystallized structure of ER-α36, and the dynamics and motions of the secondary or tertiary structure have been reported. The defining of the binding landscape of the ER-α36 LBD using the computational parameters provides a platform for understanding the drug resistance caused by the mutations. Furthermore, unravelling the mechanism of inhibition of the ER-α36 active site justifies the importance of this study. This study will significantly assist in the design of potent inhibitors to bind to the flexible LBD cavity.

Conclusively, the homology model of the ER-α36 enzyme allows us to gain insights into the active structures of the protein. The predicted protein's LBD structure shows unique features, such as the number of helices, strands, loops, and the 27 C-terminal amino acids. It also unfolds some significant similarities with its variant ER-α66. Moreover, the active site of both enzymes has about 80% conserved identity, and more than 95% of the predicted binding cavity residues occurred in the favored region. The docking binding affinity (BA) and the MMGB(PB)SA free binding energy (ΔG) of the BFB at the truncated pocket of the ER-α36 are significantly higher than those of the ER-α66 and FULV docked at the ER-α36 pocket.

Moreover, the ΔG of the BFB at the ER-α66′s pocket is significantly higher than the ΔG of the FULV at the pocket of the ER-α36. Therefore, the estrogen receptor variant (ER-α or ER-α66) can still be an explorable target for a cancer drug strategy. Although drug resistance against fulvestrant has been reported, the FDA-approved drug has been shown to have unavoidable high BA and ΔG values, indicating the probable downregulation or inhibition of the signaling activity of ER-α36.

The adapted combined computational metrics, including distance, TriCα, and dihedral angles, unfold critical motions relating to the structure, binding pocket, and mechanism of inhibition of the enzymes' functions. The dynamics and motions of the ER-α36 binding cavity have revealed flexible asymmetric opening and closing conformation. The results presented in this study on the dynamics and motion behaviors of the active conformation of the estrogen enzymes (ER-α36 and ER-α66) and the description of their binding cavity landscape will undoubtedly enhance the designing of potent structure-based compounds that will scale through the clinical trial of cancer drug development.

Author Contributions: Conceptualization: A.T.A. and S.M.; Methodology: A.T.A. and S.M.; Investigation: A.T.A. and S.M.; Data curation: A.T.A. and S.M.; Writing—Original draft preparation: A.T.A. and S.M.; Writing—Review and editing: A.T.A. and S.M. All authors have read and agreed to the published version of the manuscript.

Funding: This research received no external funding.

Institutional Review Board Statement: Not applicable.

Informed Consent Statement: Not applicable.

Data Availability Statement: Data are contained within the article.

Acknowledgments: The authors acknowledge the Centre for High-Performance Computing, Cape Town, for providing computational resources. The authors also acknowledge the University of South Africa for supporting this study.

Conflicts of Interest: The authors declare no conflict of interest.

References

1. Liu, W.J.; Zeng, X.B.; Li, J.; Zhu, K.; Lu, H.M. Comparison of the roles of estrogens and androgens in breast cancer and prostate cancer. *J. Cell. Biochem.* **2019**, *121*, 2756–2769. [CrossRef] [PubMed]
2. Zhong, L.; Li, Y.; Xiong, L.; Wang, W.; Wu, M.; Yuan, T.; Yang, W.; Tian, C.; Miao, Z.; Wang, T.; et al. Small molecules in targeted cancer therapy: Advances, challenges, and future perspectives. *Signal Transduct. Target. Ther.* **2021**, *6*, 201. [CrossRef] [PubMed]
3. Hartkopf, A.D.; Grischke, E.-M.; Brucker, S.Y. Endocrine-Resistant Breast Cancer: Mechanisms and Treatment. *Breast Care* **2020**, *15*, 347–354. [CrossRef] [PubMed]
4. Wang, X.; Zhang, H.; Chen, X. Drug resistance and combating drug resistance in cancer. *Cancer Drug Resist* **2019**, *2*, 141–160. [CrossRef] [PubMed]
5. Dimitrakakis, C.; Bondy, C. Androgens and the breast. *Breast Cancer Res.* **2009**, *11*, 212. [CrossRef]
6. Wang, Z.Y.; Zhang, X.T.; Shen, P.; Loggie, B.W.; Chang, Y.C.; Deuel, T.F. A variant of estrogen receptor-α, hER-α36: Transduction of estrogen- and antiestrogen-dependent membrane-initiated mitogenic signaling. *Proc. Natl. Acad. Sci. USA* **2006**, *103*, 9063–9068. [CrossRef]
7. Li, G.; Zhang, J.; Xu, Z.; Li, Z. ER α 36 as a Potential Therapeutic Target for Tamoxifen-Resistant Breast Cancer Cell Line through EGFR/ERK Signaling Pathway. *Cancer Manag. Res.* **2020**, *12*, 265–275. [CrossRef]
8. Pagano, M.T.; Ortona, E.; Dupuis, M.L. A Role for Estrogen Receptor alpha36 in Cancer Progression. *Front. Endocrinol.* **2020**, *11*, 506. [CrossRef]
9. Teng, L.; Su, X.; Xu, X.; Li, G.; Lin, B.; Cao, J. ER-α36: A novel biomarker and potential therapeutic target in breast cancer. *Onco. Targets. Ther.* **2014**, *7*, 1525–1533. [CrossRef]
10. Rasha, F.; Sharma, M.; Pruitt, K. Mechanisms of endocrine therapy resistance in breast cancer. *Mol. Cell. Endocrinol.* **2021**, *532*, 111322. [CrossRef]
11. Nelson, A.W.; Tilley, W.D.; Neal, D.E.; Carroll, J.S. Estrogen receptor beta in prostate cancer: Friend or foe? *Endocr. Relat. Cancer* **2014**, *21*, T219–T234. [CrossRef] [PubMed]
12. Gu, Y.; Chen, T.; López, E.; Wu, W.; Wang, X.; Cao, J.; Teng, L. The therapeutic target of estrogen receptor-alpha36 in estrogen-dependent tumors. *J. Transl. Med.* **2014**, *12*, 16. [CrossRef] [PubMed]
13. Andrahennadi, S.; Sami, A.; Haider, K.; Chalchal, H.I.; Le, D.; Ahmed, O.; Manna, M.; El-Gayed, A.; Wright, P.; Ahmed, S. Efficacy of fulvestrant in women with hormone-resistant metastatic breast cancer (mBC): A Canadian province experience. *Cancers* **2021**, *13*, 4163. [CrossRef] [PubMed]
14. Kaminska, K.; Akrap, N.; Staaf, J.; Alves, C.L.; Ehinger, A.; Ebbesson, A.; Hedenfalk, I.; Beumers, L.; Veerla, S.; Harbst, K.; et al. Distinct mechanisms of resistance to fulvestrant treatment dictate level of ER independence and selective response to CDK inhibitors in metastatic breast cancer. *Breast Cancer Res.* **2021**, *23*, 26. [CrossRef] [PubMed]
15. Bailey, S.T.; Shin, H.; Westerling, T.; Liu, X.S.; Brown, M. Estrogen receptor prevents p53-dependent apoptosis in breast cancer. *Proc. Natl. Acad. Sci. USA* **2012**, *109*, 18060–18065. [CrossRef] [PubMed]
16. Ribas, R.; Pancholi, S.; Rani, A.; Schuster, E.; Guest, S.K.; Nikitorowicz-Buniak, J.; Simigdala, N.; Thornhill, A.; Avogadri-Connors, F.; Cutler, R.E.; et al. Targeting tumour re-wiring by triple blockade of mTORC1, epidermal growth factor, and oestrogen receptor signalling pathways in endocrine-resistant breast cancer. *Breast Cancer Res.* **2018**, *20*, 44. [CrossRef] [PubMed]
17. Guo, M.; Wang, M.; Deng, H.; Zhang, X.; Wang, Z.Y. A novel anticancer agent Broussoflavonol B downregulates estrogen receptor (ER)-α36 expression and inhibits growth of ER-negative breast cancer MDA-MB-231 cells. *Eur. J. Pharmacol.* **2013**, *714*, 56–64. [CrossRef]
18. Jeong, J.H.; Jae-Ha, R. Broussoflavonol B from Broussonetia kazinoki Siebold Exerts Anti-Pancreatic Cancer Activity through Downregulating FoxM1. *Molecules* **2020**, *25*, 2328. [CrossRef]
19. Chetty, S.; Soliman, M.E.S. Possible allosteric binding site on Gyrase B, a key target for novel anti-TB drugs: Homology modelling and binding site identification using molecular dynamics simulation and binding free energy calculations. *Med. Chem. Res.* **2015**, *24*, 2055–2074. [CrossRef]
20. Morgat, A.; Lombardot, T.; Coudert, E.; Axelsen, K.; Neto, T.B.; Gehant, S.; Bansal, P.; Bolleman, J.; Gasteiger, E.; de Castro, E.; et al. Enzyme annotation in UniProtKB using Rhea. *Bioinformatics* **2020**, *36*, 1896–1901. [CrossRef]
21. Roche, D.B.; Tetchner, S.J.; McGuffin, L.J. FunFOLD: An improved automated method for the prediction of ligand binding residues using 3D models of proteins. *BMC Bioinform.* **2011**, *12*, 160. [CrossRef]
22. Wu, Y.L.; Yang, X.; Ren, Z.; McDonnell, D.P.; Norris, J.D.; Willson, T.M.; Greene, G.L. Structural basis for an unexpected mode of SERM-Mediated ER antagonism. *Mol. Cell* **2005**, *18*, 413–424. [CrossRef] [PubMed]
23. Nettles, K.W.; Bruning, J.B.; Gil, G.; O'Neill, E.E.; Nowak, J.; Hughs, A.; Kim, Y.; DeSombre, E.R.; Dilis, R.; Hanson, R.N.; et al. Structural plasticity in the oestrogen receptor ligand-binding domain. *Sci. Rep.* **2007**, *8*, 563–568. [CrossRef]
24. Zhang, B.; Kiefer, J.R.; Blake, R.A.; Chang, J.H.; Hartman, S.; Ingalla, E.R.; Kleinheinz, T.; Mody, V.; Nannini, M.; Ortwine, D.F.; et al. Unexpected equivalent potency of a constrained chromene enantiomeric pair rationalized by co-crystal structures in complex with estrogen receptor alpha. *Bioorg. Med. Chem. Lett.* **2019**, *29*, 905–911. [CrossRef] [PubMed]
25. Goujon, M.; McWilliam, H.; Li, W.; Valentin, F.; Squizzato, S.; Paern, J.; Lopez, R. A new bioinformatics analysis tools framework at EMBL-EBI. *Nucleic Acids Res.* **2010**, *38*, 695–699. [CrossRef] [PubMed]
26. Yang, Z.; Lasker, K.; Schneidman-Duhovny, D.; Webb, B.; Huang, C.C.; Pettersen, E.F.; Goddard, T.D.; Meng, E.C.; Sali, A.; Ferrin, T.E. UCSF Chimera, MODELLER, and IMP: An integrated modeling system. *J. Struct. Biol.* **2012**, *179*, 269–278. [CrossRef]

27. Tian, W.; Chen, C.; Lei, X.; Zhao, J.; Liang, J. CASTp 3.0: Computed atlas of surface topography of proteins. *Nucleic Acids Res.* **2018**, *46*, W363–W367. [CrossRef]
28. Edelsbrunner, H.; Mücke, E.P. Three-Dimensional Alpha Shapes. *ACM Trans. Graph.* **1994**, *13*, 43–72. [CrossRef]
29. Lange, A.W.; Herbert, J.M.; Albrecht, B.J.; You, Z.Q. Intrinsically smooth discretisation of Connolly's solvent-excluded molecular surface. *Mol. Phys.* **2020**, *118*, e1644384. [CrossRef]
30. Benkert, P.; Biasini, M.; Schwede, T. Toward the estimation of the absolute quality of individual protein structure models. *Bioinformatics* **2011**, *27*, 343–350. [CrossRef]
31. Wang, J.; Wolf, R.M.; Caldwell, J.W.; Kollman, P.A.; Case, D.A. Development and testing of a general amber force field. *J. Comput. Chem.* **2004**, *25*, 1157–1174. [CrossRef] [PubMed]
32. Morris, G.M.; Huey, R.; Lindstrom, W.; Sanner, M.F.; Belew, R.K.; Goodsell, D.S.; Olson, A.J. AutoDock4 and AutoDockTools4: Automated Docking with Selective Receptor Flexibility. *J. Comput. Chem.* **2009**, *30*, 2785–2791. [CrossRef]
33. Ciccone, L.; Petrarolo, G.; Barsuglia, F.; Fruchart-Gaillard, C.; Lajeunesse, E.C.; Adewumi, A.T.; Soliman, M.E.S.; La Motta, C.; Orlandini, E.; Nencetti, S. Nature-Inspired O-Benzyl Oxime-Based Derivatives as New Dual-Acting Agents Targeting Aldose Reductase and Oxidative Stress. *Biomolecules* **2022**, *12*, 448. [CrossRef] [PubMed]
34. Lee, T.S.; Cerutti, D.S.; Mermelstein, D.; Lin, C.; Legrand, S.; Giese, T.J.; Roitberg, A.; Case, D.A.; Walker, R.C.; York, D.M. GPU-Accelerated Molecular Dynamics and Free Energy Methods in Amber18: Performance Enhancements and New Features. *J. Chem. Inf. Model.* **2018**, *58*, 2043–2050. [CrossRef] [PubMed]
35. Jorgensen, W.L.; Chandrasekhar, J.; Madura, J.D.; Impey, R.W.; Klein, M.L. Comparison of simple potential functions for simulating liquid water. *J. Chem. Phys.* **1983**, *79*, 926–935. [CrossRef]
36. Ryckaert, J.P.; Ciccotti, G.; Berendsen, H.J.C. Numerical integration of the cartesian equations of motion of a system with constraints: Molecular dynamics of n-alkanes. *J. Comput. Phys.* **1977**, *23*, 327–341. [CrossRef]
37. Adewumi, A.T.; Soremekun, O.S.; Ajadi, M.B.; Soliman, M.E.S. Thompson loop: Opportunities for antitubercular demethylmenaquinone methyltransferase protein. *RSC Adv.* **2020**, *10*, 23466–23483. [CrossRef]
38. Adewumi, A.T.; Elrashedy, A.; Soremekun, O.S.; Ajadi, M.B.; Soliman, M.E.S. Weak spots inhibition in the Mycobacterium tuberculosis antigen 85C target for antitubercular drug design through selective irreversible covalent inhibitor-SER124. *J. Biomol. Struct. Dyn.* **2020**, *40*, 2934–2954. [CrossRef]
39. Post, M.; Wolf, S.; Stock, G. Principal component analysis of nonequilibrium molecular dynamics simulations. *J. Chem. Phys.* **2019**, *150*, 204110. [CrossRef]
40. Seifert, E. OriginPro 9.1: Scientific data analysis and graphing software—Software review. *J. Chem. Inf. Model.* **2014**, *54*, 1552. [CrossRef]
41. Pal, S.; Kumar, V.; Kundu, B.; Bhattacharya, D.; Preethy, N.; Reddy, M.P.; Talukdar, A. Ligand-based Pharmacophore Modeling, Virtual Screening and Molecular Docking Studies for Discovery of Potential Topoisomerase I Inhibitors. *Comput. Struct. Biotechnol. J.* **2019**, *17*, 291–310. [CrossRef] [PubMed]
42. Genheden, S.; Ryde, U. The MM/PBSA and MM/GBSA methods to estimate ligand-binding affinities. *Expert Opin. Drug Discov.* **2015**, *10*, 449–461. [CrossRef]
43. Gapsys, V.; Michielssens, S.; Peters, J.H.; de Groot, B.L.; Leonov, H. Calculation of Binding Free Energies. *Methods Mol. Biol.* **2015**, *1215*, 173–209.
44. Kang, L.; Zhang, X.; Xie, Y.; Tu, Y.; Wang, D.; Liu, Z.; Wang, Z.Y. Involvement of estrogen receptor variant ER-α36, not GPR30, in nongenomic estrogen signaling. *Mol. Endocrinol.* **2010**, *24*, 709–721. [CrossRef] [PubMed]
45. Omarjee, S.; Jacquemetton, J.; Poulard, C.; Rochel, N.; Dejaegere, A.; Chebaro, Y.; Treilleux, I.; Marangoni, E. The molecular mechanisms underlying the ER α-36-mediated signaling in breast cancer. *Oncogene* **2016**, *36*, 2503–2514. [CrossRef] [PubMed]

MDPI AG
Grosspeteranlage 5
4052 Basel
Switzerland
Tel.: +41 61 683 77 34

Biomolecules Editorial Office
E-mail: biomolecules@mdpi.com
www.mdpi.com/journal/biomolecules

www.ingramcontent.com/pod-product-compliance
Lightning Source LLC
Chambersburg PA
CBHW061610210926
51850CB00009B/2276
9783725878390